Climate Change, Circular Economy, and Industry:

Economic Approaches and Policy Implications

Seda H. Bostancı
https://orcid.org/0000-0002-3559-2224
Tekirdağ Namık Kemal University, Turkey

Seda Yıldırım
https://orcid.org/0000-0003-4367-6652
Tekirdağ Namık Kemal University, Turkey

Valentin Vasilev
https://orcid.org/0000-0002-0074-9578
Higher School of Security and Economics, Bulgaria

IGI Global
Scientific Publishing
Publishing Tomorrow's Research Today

Vice President of Editorial	Melissa Wagner
Director of Acquisitions	Mikaela Felty
Director of Book Development	Jocelynn Hessler
Production Manager	Mike Brehm
Cover Design	Jose Rosado

Published in the United States of America by
IGI Global Scientific Publishing
701 East Chocolate Avenue
Hershey, PA, 17033, USA
Tel: 717-533-8845 | Fax: 717-533-7115
Website: https://www.igi-global.com E-mail: cust@igi-global.com

Library of Congress Cataloging-in-Publication Data

LCCN: 2026007414 (CIP Data Pending)
ISBN13: 9798337396859
Isbn13Softcover: 9798337396866
EISBN13: 9798337396873

British Cataloguing in Publication Data
A Cataloguing in Publication record for this book is available from the British Library.

Editorial Advisory Board

Table of Contents

Detailed Table of Contents

The climate crisis refers to the negative consequences of human-induced economic activities on nature and human life in the 20th century. These consequences include the destruction of natural resources and land, depletion of water resources, disruption of life forms in nature, air and water pollution, loss of health, habitat loss, and displacement. Therefore, today's economic activities must consider this negative legacy of the previous century. However, this new constraint, clearly observable in nature and social life and referred to as the climate crisis, has unfortunately not yet been adequately reflected in economic theory. This situation risks facing an even bigger crisis if economic policies continue to be based on economic information. In this context, this study examines the extent to which orthodox economics and post-Keynesian economics, as a heterodox approach, have attempted to consider climate constraints from the past to the present, and reveals the extent to which economic theory is prepared for policy measures to be developed to address the climate crisis.

The sports industry, with its global reach and influence, has the potential to play a pivotal role in driving sustainability and combating climate change. This chapter explores how sports can promote sustainability through eco-friendly infrastructure, low-carbon events, and community engagement. It examines the environmental impact of sports, including carbon emissions, energy use, and waste. The chapter highlights green stadiums powered by renewable energy and the trend of organizing low-carbon, zero-waste events. It also discusses how athletes and sports organizations advocate for climate action. At the grassroots level, sports initiatives empower

This study explores how flood disasters reflect environmental injustice within the framework of risk society theory and social work. Through a comparative analysis of the 2018 Japan Floods and the 2021 Western Black Sea Flood in Turkey, it examines how modern risks disproportionately affect vulnerable populations and how disaster responses differ based on national resilience, institutional capacity, and social justice. Drawing from Ulrich Beck's risk society and Bullard's environmental justice approaches, the study reveals Japan's reflexive adaptation to disasters, emphasizing prevention, community resilience, and psychosocial recovery, while highlighting challenges in Turkey's risk preparedness, public participation, and equity in recovery. The research underscores the need for inclusive, rights-based, and socially just disaster management systems.

Access to affordable, reliable, and clean electricity is vital for economic growth, social well-being, and sustainable development. However, conventional power systems face growing challenges, with climate change posing significant risks due to their high carbon footprint. Frequent blackouts, reduced efficiency, and infrastructure damage highlight the urgent need for solutions. Power utilities must modernize infrastructure and adopt low-carbon energy sources. This chapter explores how microgrids can address these challenges by offering decentralized, sustainable power generation that enhances resilience and reliability. Microgrids support climate adaptation and mitigation efforts, ensuring energy access despite environmental uncertainties. The chapter highlights their benefits, including improved efficiency and reduced emissions, while also addressing technical, economic, and regulatory challenges that hinder widespread adoption. By leveraging microgrids, communities and industries can transition to a resilient, sustainable energy future.

Preface

INTRODUCTION

The crises arising from the immediate consequences of climate change are affecting governments and economic policies worldwide. However, experts in this field are still hesitant about making climate change a central focus in general approaches to economic theory. Although climate change is discussed in the fields of environmental economics and circular economy, there are gaps regarding the political economy of this area. The aim of this book is to discuss the connections between climate change and political economy in the policy-making process.

"Climate Change, Circular Economy, and Industry: Economic Approaches and Policy Implications" is a book that examines, discusses, and critiques how industry and other sectors are affected by climate change and the circular economy. The book also aims to draw policymakers' attention to the inseparable aspects of climate change from economic policies. Covering topics such as climate change, the circular economy, waste management, and industrial policies, this book is a valuable resource for academics, researchers, policymakers, climate researchers, and graduate students.

Coverage

The many academic areas covered in this publication include, but are not limited to:

- Political economy
- Circular economy
- Environmental economy
- Socio-Economic Perspectives
- Waste economy
- Industry
- Economic Policies
- Waste Management
- Climate Change
- Climate Action
- Corporate Management
- Energy Policies
- Energy Management
- Human Capital
- Social Work

CHAPTER OVERVIEWS

This book aims to collect international scholars' original works related with "Climate Change, Circular Economy, and Industry". This book collected 10 chapters as below:

Chapter 1: THE Climate Crisis and Orthodox and Heterodox Economics

The climate crisis refers to the negative consequences of human-induced economic activities on nature and human life in the 20th century. These consequences include the destruction of natural resources and land, depletion of water resources, disruption of life forms in nature, air and water pollution, loss of health, habitat loss, and displacement. Therefore, today's economic activities must consider this negative legacy of the previous century. However, this new constraint, clearly observable in nature and social life and referred to as the climate crisis, has unfortunately not yet been adequately reflected in economic theory. This situation risks facing an even bigger crisis if economic policies continue to be based on economic information. In this context, this study examines the extent to which orthodox economics and post-Keynesian economics, as a heterodox approach, have attempted to consider climate constraints from the past to the present, and reveals the extent to which economic theory is prepared for policy measures to be developed to address the climate crisis.

Chapter 2: Green Games: Transforming Sports into a Catalyst for Global Sustainability

The sports industry, with its global reach and influence, has the potential to play a pivotal role in driving sustainability and combating climate change. This chapter explores how sports can promote sustainability through eco-friendly infrastructure, low-carbon events, and community engagement. It examines the environmental impact of sports, including carbon emissions, energy use, and waste. The chapter highlights green stadiums powered by renewable energy and the trend of organizing low-carbon, zero-waste events. It also discusses how athletes and sports organizations advocate for climate action. At the grassroots level, sports initiatives empower communities and foster climate-conscious behaviors. Concluding with an exploration of international policies, the chapter stresses the urgent need for collaboration across sectors to embed sustainability into the sports industry, demonstrating how sports can lead the way in creating a more sustainable and resilient future for all.

Chapter 3: A Brief Study on Waste Management and Circular Policies in Tourism Industry

The tourism industry, which is a significant contributor to the acceleration of climate change and the increase in environmental pollution, must also shoulder some responsibility for mitigating this. The tourism industry is considered a comprehensive and diverse industry encompassing a wide range of activities and services. Various activities, such as transportation, dining, accommodation, and travel, as well as tourist behavior, necessitate a comprehensive examination of the tourism industry's green transformation. This study aims to conduct a specific analysis of the tourism industry regarding waste management and waste management policies.

In this study, important waste management application examples and recycling strategies will be reviewed for the tourism industry worldwide. By categorizing the main areas within the tourism industry, specific waste management strategies and practices for these subcategories and areas will be discussed. The study focuses on establishing a basic profile for waste management and recycling policies in the tourism industry.

Chapter 4: A Review of Key Indicators for Waste Management: Turkey Case

Waste management has become a prominent application with the green economy and circular economy. The most important circular application area of the 21st century can be considered as elements related to waste management processes. Innovative approaches and alternative policies are being developed for the improvement of waste management on a global scale. There may be differences between the approaches and practices adopted by each country for waste management process. However, generally, the assistance of smart technologies and the inclusion of the public in the waste management process can be considered as common policies. Technological infrastructure and integration will yield more efficient results in the long term. At this point, the aim of the study is to examine the cyclical applications in waste management. Turkey was selected as a sample country. Turkey is a developing country that has taken important steps in combating climate change. In this regard, a sample country review will be conducted by presenting useful information on waste management and circular applications.

Chapter 5: Navigating Climate Action and Sustainability-From Energy Justice to Waste Management in the Face of Climate Change

Integrating energy justice with sustainable waste management, this study constructs a comprehensive framework addressing climate change by fostering socio-economic resilience, minimizing environmental degradation, and promoting decentralized renewable energy systems through waste-to-energy technologies; employing an interdisciplinary methodology—encompassing statistical modeling, geospatial analysis, global case studies, and stakeholder interviews—the research identifies systemic barriers such as financial constraints, technological limitations, and policy fragmentation while advancing circular economy principles and participatory governance models to harmonize sustainability with social equity, ensuring transformative, inclusive, and equitable global climate resilience strategies.

Chapter 6: Problems and Trends in the Implementation of Digital Transformation of Business Models and ESG Strategies in Corporate Management

Digital transformation and ESG strategies have a key role in modern corporate management by creating opportunities for sustainable growth and competitiveness. Functional relationships and dependencies in the triad of "digitalization - sustainable development - modern management" have been analyzed. The main research thesis is that digitalization, sustainable development and successful management are systemic process phenomena that mutually arise and significantly influence each other. Some issues at the micro level have been outlined and an analysis of the interrelations between the use of digital technologies, the implementation of sustainable development goals and effective corporate governance has been conducted. With a survey conducted, which included 617 respondents from South-West region of Bulgaria, the company environment and readiness for transformation of business models were investigated, and the current state of some basic parameters related to sustainable development, corporate management and environmental and social practices.

Chapter 7: Risk Society and Environmental Justice: A Comparative Evaluation in the Context of Social Work Through the Cases of Turkey and Japan

This study explores how flood disasters reflect environmental injustice within the framework of risk society theory and social work. Through a comparative analysis of the 2018 Japan Floods and the 2021 Western Black Sea Flood in Turkey, it examines how modern risks disproportionately affect vulnerable populations and how disaster responses differ based on national resilience, institutional capacity, and social justice. Drawing from Ulrich Beck's risk society and Bullard's environmental justice approaches, the study reveals Japan's reflexive adaptation to disasters, emphasizing prevention, community resilience, and psychosocial recovery, while highlighting challenges in Turkey's risk preparedness, public participation, and equity in recovery. The research underscores the need for inclusive, rights-based, and socially just disaster management systems.

Chapter 8: The Role of Microgrids in Addressing Climate Challenges and Revolutionizing Energy Systems

Access to affordable, reliable, and clean electricity is vital for economic growth, social well-being, and sustainable development. However, conventional power systems face growing challenges, with climate change posing significant risks due to their high carbon footprint. Frequent blackouts, reduced efficiency, and infrastructure damage highlight the urgent need for solutions. Power utilities must modernize infrastructure and adopt low-carbon energy sources. This chapter explores how microgrids can address these challenges by offering decentralized, sustainable power generation that enhances resilience and reliability. Microgrids support climate adaptation and mitigation efforts, ensuring energy access despite environmental uncertainties. The chapter highlights their benefits, including improved efficiency and reduced emissions, while also addressing technical, economic, and regulatory challenges that hinder widespread adoption. By leveraging microgrids, communities and industries can transition to a resilient, sustainable energy future.

Acknowledgment

This book includes 8 chapters and gives different perspectives for Climate Change, Circular Economy, and Industry.

With the help of international scholars, both review and case studies are conducted under this book. All responsibilities of the specific chapters have belonged to their authors. Accordingly, rights, permissions and opinions of chapters are based on authors' responsibility.

Editors

Prof. Dr. Seda H. Bostancı

Prof. Dr. Seda Yıldırım

Prof. Dr. Valentin Vasilev

Chapter 1
The Climate Crisis and Orthodox and Heterodox Economics

İdris Akkuzu
https://orcid.org/0000-0003-1069-9719
Tekirdağ Namık Kemal University, Turkey

Seda H. Bostancı
https://orcid.org/0000-0002-3559-2224
Tekirdağ Namık Kemal University, Turkey

ABSTRACT

The climate crisis refers to the negative consequences of human-induced economic activities on nature and human life in the 20th century. These consequences include the destruction of natural resources and land, depletion of water resources, disruption of life forms in nature, air and water pollution, loss of health, habitat loss, and displacement. Therefore, today's economic activities must consider this negative legacy of the previous century. However, this new constraint, clearly observable in nature and social life and referred to as the climate crisis, has unfortunately not yet been adequately reflected in economic theory. This situation risks facing an even bigger crisis if economic policies continue to be based on economic information. In this context, this study examines the extent to which orthodox economics and post-Keynesian economics, as a heterodox

DOI: 10.4018/979-8-3373-9685-9.ch001

approach, have attempted to consider climate constraints from the past to the present, and reveals the extent to which economic theory is prepared for policy measures to be developed to address the climate crisis.

INTRODUCTION

The climate crisis is a socio-economic phenomenon poised to become the most significant crisis of the 21st century. This phenomenon affects not only the realm of realpolitik implemented at national and international levels, but also the field of scientific production. In this sense, the natural and social sciences, while continuing their efforts to incorporate the "climate" variable into scientific models at a theoretical level, are also engaged in discussing the "climate crisis" phenomenon, explaining its current and potential social impacts, and developing mechanisms to prevent them. On the other hand, it is observed that economics, in this process, is still quite shy in addressing the climate issue with the seriousness and clarity it deserves, and in expressing the limitations of orthodox economics in doing so and having the courage to overcome these limitations. Indeed, orthodox economics limits its understanding of the climate crisis to a narrow and limited framework of "externalities" and "market failures," ultimately proposing market-based solutions such as carbon pricing. However, these proposals are only a part of the search for optimality, focusing only on the aspect that falls within the realm of efficient market thinking, rather than viewing the issue holistically (Keen, 2021). This is one of the reasons behind the intense interest shown by heterodox economics in the climate crisis in recent years. Heterodox economics, which attempts to understand the nature of things outside the mainstream economic science world, is not actually a homogeneous whole. It points to a heterogeneous structure encompassing different approaches to economics, political economy, socio-economics, ecological economics, and behavioral economics (Spash, 2020; Storm, 2021). One of these approaches is Post-Keynesian economics. Post-Keynesian economics claims to offer a more systemic, holistic, and realistic framework with concepts such as uncertainty, demand-side growth, endogenous money theory, and financial instability (Fontana, & Sawyer, 2016; Politt, 2019). This article examines the

Post-Keynesian approach to the climate crisis through the lens of growth imperative, public investment, green finance, and inequality, and explores the macroeconomic models it proposes to address the shortcomings of neoclassical models regarding climate issues (Klitgaard, & Krall, 2012; Rezai, et al., 2013; Kemp-Benedict, 2018).

The Climate Problem and the Silence/ Inadequacy of Neo-Classical Economics

Although climate change is one of the most serious problems threatening the existence of humanity, mainstream economics has long avoided paying attention to it. Indeed, Oswald and Stern (2019, p. 10) argue that the number of publications on the "climate crisis" in leading academic economics journals worldwide is disproportionately low compared to the magnitude and importance of the problem. Since we have no reason to believe that there are more important issues than climate change on the research agenda, the answer to this lack of attention must be sought in the methodological approach that mainstream economics has taken towards the subject. For over a century, neoclassical economics, which has been at the helm of academic economics, has shown, as will be detailed below, that the current state of economic theory lags behind the global economic policy-making processes, even when considering only the 2015 Paris Climate Agreement (United Nations, 2015) and the reports of the United Nations Intergovernmental Panel on Climate Change (IPCC).

"The IPCC is an intergovernmental body open to all member countries of UNEP (United Nations Environment Programme) and WMO (World Meteorological Organization). The IPCC has three working groups and a Task Force on National Greenhouse Gas Inventories. Working Group I assesses the scientific aspects of the climate system and climate change. Working Group II addresses the vulnerability of socio-economic and natural systems to climate change, the negative and positive consequences of climate change, and adaptation options. Working Group III assesses options for limiting greenhouse gas emissions and mitigating climate change in other ways. The IPCC's First Assessment Report was published in 1990 and confirmed the scientific basis of concerns about climate change" (Reinman, 2012).

The reason for this widening gap between theory and policy stems not only from the issue of climate change, but from the methodological limitations that mainstream economics has faced since its inception, in incorporating changing economic, social, and environmental conditions. A general overview of these limitations is beyond the scope of this essay. Instead, it would be more appropriate to list the fundamental propositions of neoclassical economics concerning climate change and to explain the concepts that define the limitations of these propositions.

It should be noted first that neoclassical economics, as a model that envisions unlimited economic growth, has not modelled natural elements, including climate, which it categorizes as resources, as constraints; rather, it has accepted the unlimited nature of resource use in terms of growth and prosperity. Here, the assumptions of substitutability among resources and the idea that nature is an externality in terms of the market mechanism have formed the main outlines of the research and proposals (Masini, 2021). Accordingly, the transformation of human and physical capital brought about by technological developments over time can substitute natural resources; and effective market outcomes can be achieved through market interventions characterized by "internalization of externalities." However, in the classical political economy period preceding neoclassical economics, there was a search for an understanding of whether growth had a constraint between land rent and environmental factors (Galbraith, 2021), and in this respect, it tried to adopt a more realistic approach than neoclassical economics.

However, the Cobb-Douglas-esque neoclassical production functions developed later are based on the assumption of continuous substitution among factors of production. This leads to the misconception that when natural capital is depleted, it can be replaced by physical capital. In other words, the neoclassical economic approach to the climate crisis is that natural resource constraints can be overcome through technological progress and market mechanisms. According to this view, when natural capital (e.g., a clean atmosphere or biodiversity) is depleted, the price mechanism will kick in, and firms will substitute this resource with "physical capital" (machinery, technology). In neoclassical economics, technological change is generally included in models as an "exogenous" variable. Technology, in a sense, is a phenomenon that develops spontaneously outside the economic sphere. This assumption assigns a foreign and salvific, yet misleading, role to technology in solving the climate crisis.

One of the significant steps that challenged the notion of a problem-free relationship between nature and growth was the 1972 publication of *The Limits to Economic Growth* by Meadows et al., thinkers from the Club of Rome. This work, on the one hand, reopened the discussion on the scarcity phenomenon in the relationship between growth and population, bringing pessimistic views back into mainstream thought. More importantly, it warned that renewable natural resources could not keep pace with the current rate of economic growth and consumption, thus leading to a serious resource crisis in the near future (p. 55).

This work, coinciding with the oil shocks of the 1970s and the beginning of the restructuring process of the world economy that would later be called globalization, unfortunately only stimulated Malthusian debates in economic thought in two directions, but failed to ignite a signal for the development of a radical, nature-climate-based approach in mainstream theory. On the contrary, as a rejection of this movement, it became the motivating factor for another orthodox effort that has shaped the approach of economic theory to the climate crisis from those years to the present day. The economist Nordhaus, who was the author of this effort and was therefore awarded the Nobel Prize in 2018, and his ideas have been accepted as the prevailing currency in the field and, with some minor revisions, have continued to exist until today.

In 1992, Nordhaus developed the Dynamic Integrated Model of Climate and the Economy (DICE) as a climate-based macroeconomic model. With this model, he aimed to reveal the economic impacts of climate change, the theoretical avenues for reducing the costs it imposes on the economy, and the corresponding economic policies to be developed (Arbues, 2021, p. 11). As a kind of integrated assessment model, the DICE model aimed to express the social cost of carbon by establishing an optimization between climate and the economy and attempting to find the marginal cost of an additional unit of greenhouse gas emissions from economic activities. Thus, an optimal carbon tax application to mitigate the damage would be possible (Nordhaus, 1993). In a way, this meant bringing together the mainstream economic proposals, which were confined to the pricing and limitation of greenhouse gas emissions in the climate-economy relationship, under the umbrella of a carbon tax (Galbraith, 2021, p. 1119).

The DICE model (Nordhaus, 2019), which appears to be limited in its attempt to account for the variability of climate and non-climate economic factors in long-term growth projections, still serves as a reference source for neoclassical economic theory and continues to produce optimistic information that does not share the pessimism held by climate policy circles.

In their 2022 study, Stern and Stiglitz highlighted the need for new approaches to the relationship between climate change and economic theory, revealing the shortcomings of such optimization-oriented models (Stern and Stiglitz, 2022). These models disregard "uncertainty," treating risks as calculable and manageable phenomena (p. 182), whereas the climate system contains irreversible "tipping points," such as the melting of glaciers, making cost-benefit analysis impossible (p. 191). This is because at each tipping point, the cost deviates from the pattern established in the previous function, following an unpredictable trajectory. According to the authors, another problem is the assumption that efficient market conditions can be achieved solely through a carbon tax. However, markets, encompassing areas such as finance, technology, and information, possess the potential to produce unsuccessful outcomes in a multifaceted way (p. 197).

Another of the most heated debates regarding climate revolves around the proposed discount for calculating the present value of future damages. This debate ranges from Nordhaus's suggestion of a high discount to Stern's (2007) insistence on a low discount rate, thus aiming to develop climate policies that focus on the present rather than the future. A high discount, in a sense, is based on the assumption that future generations will have a better standard of living than today, but it reduces the present value of future damages to very low amounts, thus arguing against the inefficiency of large-scale investments based on climate change (Stern and Stiglitz, p. 196).

Post-Keynesian Economics and the Climate Problem

The observation made above regarding neoclassical economics' long-standing indifference to the climate crisis can also be applied to Post-

Keynesian economics. Post-Keynesian economics, almost exclusively focused on maintaining capital accumulation and full employment, has largely ignored climatic and environmental problems. In recent years, this approach has begun to be abandoned, and efforts to place the relationship between ecology and economics on a more solid ontological foundation through concepts such as the monetary nature of production, "path dependence," and "fundamental uncertainty" have visibly increased. Accordingly, Post-Keynesian Economics, unlike neoclassical approaches, does not view money merely as a medium of exchange or a veil over real variables, but rather as the starting and ending point of the production process. In this sense, the endogenetic nature of money in Post-Keynesian economics provides a crucial explanatory framework for the financing opportunities needed for climate policies.

As is known, a typical production process is formed by collecting savings and directing them towards investment. At this stage, Post-Keynesian Economics, with its endogenous money argument, emphasizes the power of financial markets to create financing by loosening the dependence of investments on savings. Accordingly, firms begin their production activity with initial financing, which is usually met through loan requests from banks. In this process, banks create deposit money simply by issuing loans to the market, and this process is called "endogenous money". At this stage, financial markets select investment projects they deem worthy of credit, effectively acting as a critical "selective" jury in determining the production structure through new investments. Indeed, assuming that financial markets consider fossil fuel projects "profitable" and climate-based projects "risky," the financial system, solely due to its ontological position, deepens the carbon lock-in crisis in the economy. At the end of the production process, firms repay their debts with the income they earn from selling the products they produced thanks to the initial financing ("final financing"). This cycle, unless there is external intervention, risks continuously producing results that will fuel the climate crisis. Therefore, if the aim is to expand real economic activities that incorporate the climate issue into the equation, it is necessary to actively intervene and redirect the capacity of financial markets to create credit and direct the real economy, shifting them from carbon-intensive sectors to carbon-reducing/carbon-free sectors. In fact, the transition to a low-carbon economy through the redirection of financial markets, independent of the savings enabled by the internal money supply, can be hampered in developing economies with their current foreign exchange debt/deficit (Löscher and Kaltenbrunner, 2023).

As is known, investments do not only serve the purpose of increasing wholesale demand in the macroeconomy. They also determine future supply and production capacity and the direction of technological development through additions to the existing capital stock. In this sense, according to the Post-Keynesian approach, the economy may not spontaneously move towards a predetermined equilibrium path (steady state). On the contrary, past investment decisions carry the potential to restrict current path options (path dependence). Indeed, an ongoing investment in a coal-fired power plant is creating a crisis in infrastructure that will generate emissions for decades to come, thus blocking any potential return on investment. Therefore, the composition and content of investments, whether carbon-based or anti-carbon, are at least as important as the total size of the investments.

On the other hand, investment decisions are actually made under "fundamental uncertainty" and are guided by "animal spirits." This fundamental uncertainty, stemming from the physical and transition risks created by climate change/crisis, could lead firms and financial markets to avoid long-term green investments by increasing their liquidity preferences. This situation necessitates the intervention of governments and central banks as "market makers" or "risk takers" to reduce uncertainty.

Another controversial issue in economic theory and the climate problem concerns the growth imperative. Even in the Post-Keynesian approach, there is an ongoing debate about the compatibility/incompatibility of capitalism with ecological limits. This debate oscillates between the possibilities of "growth imperative" and "stagnation" (Huwe and Rehm, 2022, p. 400). The central argument in this debate is the need for continuous capital accumulation due to capitalism's inherent drive for competition and profit. Here, there is a general disagreement about whether nature constitutes a limit to the pace of growth. Mainstream economics generally argues that environmental factors and economic growth patterns can be decoupled and that a green economy transition will have no long-term impact on the pace of growth. Post-Keynesian economics, however, approaches the assumption that a market-oriented economic ontology can simultaneously provide sustainable resources and full employment with skepticism (Heise, 2022).

In other words, the proposition that an economy can function effectively even at zero or negative growth rates under conditions of a demand-driven and financialized production economy is debatable.

In the policy arena, the question of achieving an optimal climate change target and doing so within a specific economic growth trajectory is therefore an invalid pursuit for most Post-Keynesian economists. They predict that climate-based, sustainability policies will lead to stagnation rather than growth in a financialized economy, and that the resulting employment problems and subsequent corporate and household debt crises cannot be solved by simply leaving the market to its own devices. In such a situation, public intervention will again be necessary to ensure wealth distribution and employment growth (Heise, 2022).

Therefore, according to Post-Keynesian economics, there is a conflict of interests between sustainability and growth, between firms' profit motive and competitive pressure, and ecological balance and societal well-being. This makes it impossible to achieve both goals simultaneously and through separate means. In this respect, policy mechanisms are needed to guide the market towards climate-based investment decisions even at the investment decision-making stage of firms. If this stage is not given sufficient importance, leaving the market to overcome climate constraints solely through taxation risks causing a recession. This, as stated above, can lead to an economic and financial crisis. This situation can only be overcome by the state, this time through distribution and employment policies.

Huwe and Rehm link the formation of the climate problem, both in terms of its causes and consequences, to class and global inequalities (2022, p. 399). According to them, carbon emissions are distributed very unevenly among different income groups. The carbon footprint of the high-income group is many times greater than that of the poor, both in absolute terms and per capita. For example, in the EU, the footprint of the richest 1% is 22 times higher than the sustainable level. While the consumption of the wealthy is focused more on carbon-intensive and elastic goods such as transportation, the consumption of the poor is based on basic needs such as heating and food.

From another perspective, since the marginal propensity to consume of low-income groups is relatively higher, a redistribution of social income from profit to wages (or from rich to poor) would increase aggregate demand and thus positively impact economic growth. However, this could again lead to a potential dilemma between growth and sustainability. Huwe and Rehm (2022) suggest that this problem can be overcome by considering the different consumption tendencies of different groups. According to them, the consumption expenditures of low-income groups (housing, basic services) are directed towards areas that are easier to reduce in terms of carbon footprint. In this respect, efforts to tax or restrict luxury consumption by high-income groups would contribute to a reduction in total emissions. In short, Post-Keynesian economics places the issue of income distribution at the center of climate and economic analysis and brings policy recommendations regarding unequal distribution to the ecological level.

These views on the parameters behind the neoclassical economics' limited relationship with climate actually stem from the general assumptions of Post-Keynesian economics regarding the nature of economic activity and the acceptance of human behavior; they do not represent a comprehensive and well-rounded theoretical proposal. As stated at the beginning of this study, Post-Keynesian economics, among heterodox economic approaches, has been hesitant to address the climate problem, seeking solutions through hypotheses such as demand-side growth, endogenous money, uncertainties, and financialization. Nevertheless, in recent years, efforts to develop a holistic theoretical framework have been observed.

One of these efforts is by Dafermos et al. (2018), who, using the SFF macroeconomic model, demonstrated the negative impact of climate change on financial stability and, consequently, on real economic activity, and underlined the importance of "green" monetary policy along with concrete tools to mitigate this negativity. One of the key features that distinguishes this study is that it offers a concrete and measurable quantitative analysis of the relationship between climate and the financial system by creating a kind of computational structure. Approaches to measuring resources with material units, such as Material Flow Analysis or Ecologically Unequal Exchange, which have been used for a long time in ecological economics, have been attempted to be made possible for the hypothetical world of Post-Keynesian economics. Accordingly, the study establishes a cyclical exchange system between the real economy and the financial sector, and attempts to address the flow of physical matter such as energy and emissions within this system using the same model. The model states that economic activities, operating with familiar logic, produce emissions, these emissions lead to an increase in average temperature, and this temperature increase negatively affects both physical capital and labor productivity through "damage functions" (pp. 219, 222).

In fact, this approach, which attempts to overcome the dichotomy created between climate and the economy in mainstream economics by showing the relationship between the climate crisis, financial markets, and the real sector in a single model, emphasizes that the climate problem, the increase in average temperature, is not only a climatic problem but also a systemic financial risk that will paralyze economic activity. The model shows that climate-related shocks erode the capital stock in the economy, negatively affect profitability, lead to a decline in financial capacity as profitability decreases, and cause financial assets to lose value, carrying the risk of heading towards a systemic financial crisis (pp. 227-228).

The relationality described here is one not anticipated in mainstream economics. As stated in the previous section, orthodox predictions of the climate problem, specifically the increase in average temperature, fall considerably short of the expectations of global policy circles; on the other hand, it is argued that there is no reason for the current supply of renewable resources to constitute a constraint on the necessary growth rate. These two assumptions naturally lead to the prediction that the climate variable is not of a magnitude that could cause a systemic crisis in the financial and real economy. In this case, mainstream economics deals with the climate problem in a one-sided way, that is, focusing on slowing down the pace of climate change caused by economic activities and reducing environmental damage.

Post-Keynesian economics differs in this respect, addressing the climate-economy connection in a more realistic way. Assuming a two-way interaction and that the logic of the market alone does not have the capacity to eliminate this interaction, it introduces policy proposals not found in mainstream economics. One of these policy measures is Green Central Banking. Central banks, in order to support the green economy, have the opportunity to increase the value of green bonds by issuing them and creating demand for them, thereby increasing the financial strength of economic sectors that will engage in green-based activities. Green investments that initially seem unprofitable become profitable in this way, and the proliferation of such investments is ensured. In short, thanks to green monetary policy supported by other financial and structural policies, the path to decarbonization can be opened (pp. 230-232).

The Post-Keynesian economic policy framework for the climate crisis aims not only to correct market failures but also to restructure market functioning, investment decisions, and institutions while considering "public interest" and "ecological limits." This approach was introduced into the literature by Whalen (2023) as a combination of Hyman Minsky's concept of "containing mechanisms" and William Kapp's analysis of "social costs." Accordingly, environmental pollution should be viewed not as an "externality" or an exceptional market failure, but as an inevitable consequence of profit-driven production. Minsky's hypothesis that financial instability is inherent in capitalism overlaps with Kapp's argument that environmental degradation is inherent in the market economy. Economic actors, chasing increases in the value of financial assets and attempting to do so through short-term lenses, accelerate ecological destruction by pursuing the relatively high value of assets outside the green economy. This hinders long-term green investments and encourages firms to shift environmental costs onto society. At this stage, "restrictive mechanisms" are crucial to prevent instability. These include not only market tools like carbon pricing, but also direct quantitative restrictions, technology standards, bans, and strict science-based regulations.

An example of such fundamental measures is the "Global Green New Deal" proposal. According to Pollin (2016), annual green investments of approximately 1.5-2% of global GDP are needed. This would allow for a reduction of around 40% in global emissions over a 20-year period. During the transition, the balance between climate, economy, and social structure should be achieved through measures such as income support and job guarantees for job losses in fossil fuel-dependent industries.

CONCLUSION

Neoclassical economics' optimization-focused, equilibrium-assumption-based models, which exclude uncertainty and financialization, are insufficient in the face of the urgency and complexity of the climate problem. The neoclassical approach to the climate crisis is based on the assumption that the market mechanism can overcome natural resource and ecological constraints through technological development and efficient operation. The underlying assumption of resource substitution, and the prediction that the future degree of global warming will not negatively impact the current growth rate, are questionable along with other propositions.

In contrast, Post-Keynesian Economics attempts to offer a more realistic analytical framework that focuses on uncertainty, the endogenous nature of money, demand-driven growth, and income distribution issues, demonstrating the interconnections between climate, the real economy, the financial economy, and social structures. In this sense, combating the climate crisis is not simply about correcting failing markets and developing policies to prevent climate degradation by isolating climate from the economy. On the contrary, it requires achieving a fundamental and structural transformation in the economy. To achieve this transformation, it is necessary to abandon the assumption of economic growth necessity and the climate-economy dichotomy; to increase green public investments at local, national, and global levels; to restructure financial markets to support green investments, primarily through Green Central Banking; and to implement transition mechanisms that will reduce existing ecological and social inequalities among different income groups.

REFERENCES

Arbues, I. C. (2021). *Understanding the feedbacks between climate change and the global economy: an ecological, post-Keynesian modelling approach for the European Green Deal*, Unpublished Master thesis, Delft University of Technology.

Dafermos, Y., Nikolaidi, M., & Galanis, G. (2018). Climate change, financial stability and monetary policy. *Ecological Economics*, *152*, 219–234. DOI: 10.1016/j.ecolecon.2018.05.011

Fontana, G., & Sawyer, M. (2016). Towards post-Keynesian ecological macroeconomics. *Ecological Economics*, *121*, 186–195. DOI: 10.1016/j.ecolecon.2015.03.017

Galbraith, J. K. (2021). Economics and the climate catastrophe. *Globalizations*, *18*(7), 1117–1122. DOI: 10.1080/14747731.2020.1807858

Heise, A. (2022). *Growth and sustainability in post- Keynesian perspective: Some notes*. ZÖSS-Discussion Papers.

Huwe, D., & Rehm, M. (2022). The ecological crisis and post-Keynesian economics – bridging the gap? *European Journal of Economics and Economic Policies: Intervention*, *19*(3), 397–414.

Keen, S. (2021). The appallingly bad neoclassical economics of climate change. *Globalizations*, *18*(7), 1149–1177. DOI: 10.1080/14747731.2020.1807856

Kemp-Benedict, E. (2018). Energy, growth, and sustainability: Five propositions. *Ecological Economics*, *154*, 78–83.

Klitgaard, K., & Krall, L. (2012). Ecological economics, degrowth, and institutional change. *Ecological Economics*, *84*, 247–253. DOI: 10.1016/j.ecolecon.2011.11.008

Löscher, A., & Kaltenbrunner, A. (2023). Climate change and macroeconomic policy space in developing and emerging economies. *Journal of Post Keynesian Economics*, *46*(1), 113–141. DOI: 10.1080/01603477.2022.2084630

Masini, F. (2021). William Nordhaus: A disputable Nobel [Prize]? Externalities, climate change, and governmental action. *European Journal of the History of Economic Thought*, *28*(6), 985–1004. DOI: 10.1080/09672567.2021.1963798

Meadows, D. H., Meadows, D. L., Randers, J., & Behrens, W. W. (1972). *The Limits to Growth*. Universe Books.

Nordhaus, W. (1993). Reflections on the economics of climate *change*. *The Journal of Economic Perspectives*, *7*(4), 11–25. DOI: 10.1257/jep.7.4.11

Nordhaus, W. (2019). Climate change: The ultimate challenge for economics. *The American Economic Review*, *109*(6), 1991–2014. DOI: 10.1257/aer.109.6.1991

Oswald, A. J., & Stern, N. (2019). *Why does the economics of climate change matter so much, and why has the engagement of economists been so weak?* https://www.andrewoswald.com/docs/ClimatechangeOswaldSternSept2019forRES.pdf

Politt, H. (2019). *The contribution of post-Keynesian economics to climate policy and meeting global decarbonisation targets*, https://www.postkeynesian.net/downloads/events/Pollitt_2019.pdf

. Pollin, R. (2016). *Global Green Growth for Human Development*, UNDP Human Development Report.

Reinman, S. L. (2012). Intergovernmental panel on climate change (IPCC). *Reference Reviews*, *26*(2), 41–42. DOI: 10.1108/09504121211205250

Rezai, A., Taylor, L., & Foley, D. (2013). Climate change and macroeconomic structural change. *Cambridge Journal of Economics*, *37*(5), 1207–1226.

Spash, C. L. (2020). A tale of three paradigms: Realising the revolutionary potential of ecological economics. *Ecological Economics*, *169*, 1–14. DOI: 10.1016/j.ecolecon.2019.106518

Stern, N. (2007). *The economics of climate change: The Stern review*. Cambridge University Press. DOI: 10.1017/CBO9780511817434

Stern, N., Stiglitz, J., & Taylor, C. (2022). The economics of immense risk, urgent action and radical change: Towards new approaches to the economics of climate change. *Journal of Economic Methodology*, *29*(3), 181–216. DOI: 10.1080/1350178X.2022.2040740

Storm, S. (2021). The political economy of decarbonization: A global perspective. *Development and Change*, *52*(4), 949–970.

United Nations. (2015). Paris Agreement. In report of the conference of the parties to the United Nations Framework Convention on Climate Change (21st Session, 2015: Paris).

Whalen, C. J. (2023). Minsky meets Kapp: A post-Keynesian institutionalist approach to addressing climate change. *Journal of Economic Issues*, *57*(2), 598–605. DOI: 10.1080/00213624.2023.2202560

Chapter 2
Green Games:
Transforming Sports Into a Catalyst for Global Sustainability

Amitava Pal
https://orcid.org/0009-0007-8934-2718
ICFAI University Jharkhand, India & International Institute of Sports and Management, India

M. Rajkumar
https://orcid.org/0000-0002-7606-0152
ICFAI University, Jharkhand, India

Kavita Mathad
https://orcid.org/0000-0002-0058-9084
Presidency University, Bangalore, India

ABSTRACT

The sports industry, with its global reach and influence, has the potential to play a pivotal role in driving sustainability and combating climate change. This chapter explores how sports can promote sustainability through eco-friendly infrastructure, low-carbon events, and community engagement. It examines the environmental impact of sports, including carbon emissions, energy use, and waste. The chapter highlights green stadiums powered

DOI: 10.4018/979-8-3373-9685-9.ch002

by renewable energy and the trend of organizing low-carbon, zero-waste events. It also discusses how athletes and sports organizations advocate for climate action. At the grassroots level, sports initiatives empower communities and foster climate-conscious behaviors. Concluding with an exploration of international policies, the chapter stresses the urgent need for collaboration across sectors to embed sustainability into the sports industry, demonstrating how sports can lead the way in creating a more sustainable and resilient future for all.

INTRODUCTION

Sustainability has emerged as a critical focus across sectors worldwide, driven by the urgent need to address climate change and environmental challenges (Gamage et al., 2024). The sports industry, known for its cultural significance and global reach, faces mounting pressure to align with these sustainability goals. Events like the Olympics and FIFA World Cup highlight the significant environmental footprint of sports, including energy consumption, carbon emissions, and waste generation (Geneva Environment Network, 2023). However, sports also offer immense potential to inspire change. By adopting sustainable practices, the industry can leverage its influence to promote environmental responsibility, making it a key player in global sustainability efforts (Abbass et al., 2022).

With its massive global reach and influence, the sports industry has a significant environmental footprint. From energy-intensive stadiums to the carbon emissions generated by travel and waste produced during significant events, sports have long contributed to environmental degradation. However, as the urgency of climate change grows, the sports sector increasingly recognizes the need to adopt sustainable practices (Lokesh Bohra, 2024). This chapter explores how the sports industry can mitigate its environmental impact and become a powerful driver of global sustainability. By examining key areas such as carbon emissions, waste management, and renewable energy, we will uncover the challenges and opportunities for transformation, showing how sports can play a crucial role in the fight against climate change.

CARBON EMISSIONS IN THE SPORTS INDUSTRY

The sports industry is responsible for significant carbon emissions that contribute to global warming and climate change (Wilby et al., 2022). Carbon emissions primarily come from three primary sources: energy consumption and materials used in sports infrastructure, travel, and waste generated.

Climate Change Risks and the Sports Industry

The carbon emissions produced by the global sports industry contribute approximately 0.6% of worldwide greenhouse gas emissions, comparable to the emissions of a medium-sized country . These emissions contribute to climate change risks such as rising temperatures, melting ice caps, and more frequent extreme weather events. Such changes can disrupt sports activities, damage infrastructure, and impact athlete performance. Addressing these risks requires minimizing carbon footprints through sustainable practices in travel, infrastructure, and energy use.

- **Energy Consumption and Greenhouse Gas Emissions:** Stadiums, arenas, and sports facilities use a lot of energy for lighting, heating, cooling, and powering electronic systems like scoreboards and big screens (International Energy Agency, 2024). For example, the energy required to keep a large stadium lit for a football game can be equivalent to the energy thousands of households use yearly. Most of this energy comes from non-renewable sources like coal or natural gas, which release greenhouse gases (GHGs) that harm the environment.
- **Resource Consumption in Large-Scale Events:** Large-scale sports often involve using vast resources such as water and energy. For example, maintaining a lush green pitch for football requires vast amounts of water, particularly in dry regions. At the same time, the materials needed for food, merchandising, and souvenirs also contribute to excessive resource use. Imagine a large tennis tournament held on an outdoor court. Every time it rains, water is used to clean the surface, and a significant amount of energy is needed to

dry the court for players. The event's environmental footprint will increase significantly if the tournament organizers do not focus on water conservation and energy-efficient methods.

- **Travel, Infrastructure, and Consumption:** Sports events, especially global ones like the Olympics, World Cup, or international tennis tournaments, often require teams, athletes, and thousands of spectators to travel long distances by planes, cars, and buses. Air travel, in particular, is a major contributor to carbon emissions. For example, flying athletes and fans from different countries to a single location for the FIFA World Cup generates tons of carbon emissions. The construction and maintenance of sports infrastructure also play a role. Building new stadiums, arenas, and other sports venues often involves heavy use of materials like steel, cement, and glass, which have a high carbon footprint. The Paris 2024 Olympics is committed to halving the Games' carbon footprint compared to London 2012 and Rio 2016, targeting 3.5 million tonnes of CO_2 equivalent. This innovative approach shifts from post-games carbon assessments to pre-games planning and monitoring, embedding sustainability into every aspect of the organization. Key initiatives include constructing only one new venue, the Aquatic Centre, using low-carbon materials, connecting venues to the public electricity grid powered by 100% renewable energy, and offering twice as many plant-based meal options to cut the carbon footprint of meals by 50%. Mobility solutions include promoting cycling with 20,000 temporary bike parking spaces and a 415 km cycling lane network. Additionally, Paris 2024 funds carbon avoidance and capture projects, including reforestation in France and global initiatives aligned with its health, education, and biodiversity protection values. While no longer using "carbon neutrality," the Games uphold a strong commitment to meaningful climate action and a sustainable legacy.

Waste Generation and Resource Use

If not adequately managed, sports events and facilities can generate significant waste, contributing to environmental degradation. Large waste is generated at major sports events, such as the Super Bowl or the

Olympics. This includes single-use plastics (bottles, food containers, and packaging material), food waste from concession stands, and general waste from thousands of spectators (Naylor, 2025). A significant sports event can produce tons of plastic waste in a single day, which, if not correctly disposed of or recycled, can end up in landfills or pollute rivers, oceans, and ecosystems. According to FIFA, 2022 the FIFA World Cup Qatar 2022™ showcased a strong commitment to sustainable waste management by minimizing waste generation and maximizing recycling efforts. During stadium construction, 79% of waste was diverted from landfills, while 77% of operational waste was recycled or composted during the tournament. Initiatives like the Sustainable Sourcing Code, compostable packaging, reusable service ware, and recycling-friendly signage reduced waste significantly. Food waste was minimized through targeted policies, with 133 food donation pickups aiding local communities. Awareness campaigns, led by FIFA Legend Marcel Desailly, emphasized recycling. Overall, 67% of waste across all sites was recycled or composted, exceeding Qatar's national targets, with residual waste converted to energy, setting a benchmark for sustainability in global sports events.

KEY CHALLENGES AND OPPORTUNITIES FOR CHANGE

While the environmental impact of the sports industry is undeniable, there are several key areas for improvement, as well as opportunities for innovation that can help reduce this footprint.

- **Travel Emissions:** One of the biggest challenges in reducing the carbon footprint of sports is the need for travel, particularly air travel. Athletes, teams, and spectators often need to travel long distances to participate in events, which results in high carbon emissions. As per Sustainable Travel International, 2024, overall tourism accounts for approximately 8% of global carbon emissions, with contributions coming from activities such as air travel, boat rides, accommodation, and the production of souvenirs

- **Waste Generation:** Another challenge is the massive waste generated during large-scale events. Single-use plastics, food packaging, and merchandise packaging significantly contribute to environmental damage.
- **Energy Use in Stadiums:** Many sports facilities still rely on non-renewable energy sources that contribute to GHG emissions. Although some stadiums are adopting solar panels or other renewable energy sources, much of the industry still relies on energy that harms the environment. As per Khalid & Rogstam, 2013, the average annual energy consumption of a modern medium-sized stadium with a seating capacity of 55,000 is approximately 10,000 MWh, corresponding to 3,600 metric tons of CO2 equivalent emissions.

OPPORTUNITIES FOR CHANGE

Sustainability efforts in large-scale events focus on promoting eco-friendly travel options like carpooling, public transportation, and shuttle services to reduce carbon emissions. Waste reduction initiatives, including recycling, composting, and repurposing materials, aim to minimize landfill contributions. Energy-efficient infrastructure, such as LED lighting and renewable energy sources like solar panels and wind turbines, highlights the growing commitment to reducing environmental impact while maintaining operational efficiency.

To address travel-related emissions, sports organizations can encourage sustainable transportation, such as electric buses, trains, or carpooling. The Tokyo 2020 Olympics prioritized sustainability through innovative measures. To reduce emissions, over 90% of the transportation fleet, including buses, cars, and trucks, comprised electric vehicles (EVs). The city also expanded its public transportation network, encouraging spectators to opt for eco-friendly travel. Venues were powered by renewable energy sources like solar, wind, and geothermal power, contributing to the Games' goal of carbon neutrality. The remaining emissions were offset through global projects such as reforestation and renewable energy initiatives. Public awareness campaigns and partnerships with environmental organizations also educated participants and spectators on waste reduction, energy conservation, and sustainable living, promoting sustainability beyond the event itself.

Many sports organizations are working on reducing waste by switching to reusable or compostable products. For instance, due to a robust waste and recycling program, the FIFA World Cup 2018 in Russia has been praised as a hosting success, with clean stadiums, Fan Fests, and public areas. The Waste Management Concept introduced digital communication, recyclable materials, and color-coded bins for waste segregation. Guidelines ensured effective waste handling and monitoring. Japanese fans set an example by tidying up after games and inspiring others. With hopes of reducing waste further in future tournaments, this year's efforts may serve as a blueprint for sustainable waste management at large-scale events (International Environmental Technology, 2018)

More recently, the NFL has made significant strides in addressing its environmental impact, particularly during major events like the Super Bowl. Programs like NFL Green aim to create lasting green legacies in host communities by collaborating with sponsors, local committees, and nonprofits. Initiatives include renewable energy adoption, with stadiums like Lincoln Financial Field and Gillette Stadium incorporating solar panels and wind turbines to offset energy consumption. The league purchases Renewable Energy Certificates (RECs) to match energy usage at events, promoting green electricity capacity. Waste management efforts involve recycling, composting, and e-waste recycling drives, while water conservation measures are implemented to address significant resource demands (Schaal, 2024). NFL Green also funds community greening projects, including tree planting, habitat restoration, and urban gardens, fostering sustainable practices. The league's commitment to becoming a "net zero" emissions organization involves reducing its carbon footprint through efficient travel and supporting CO2 offsetting projects. These combined efforts demonstrate the NFL's dedication to making its events environmentally sustainable while engaging fans and communities in conservation efforts (Schaal, 2024).

Regarding energy-efficient infrastructure, Mercedes-Benz Stadium in Atlanta is the first stadium worldwide to achieve the TRUE Platinum Certification for zero waste. It diverts over 90% of waste from landfills through recycling and composting, with zero waste stations placed throughout the stadium. The stadium partners with organizations like Coca-Cola and Georgia Power to enhance sustainability efforts. It features 4,000 solar PV panels producing 1.6 million kilowatt hours of renewable energy annually and a 2.1 million gallon stormwater management system. The stadium has reduced water use by 47% and promotes energy efficiency, achieving LEED Platinum certification. Public transport options, EV charging stations, and sustainable urban gardens further support its eco-friendly initiatives. Additionally, the stadium runs a STEAM curriculum for students in grades 3-12, connecting education with sustainability practices (Mercedes-Benz Stadium, 2024).

Green Energy Policies and Waste Management

Adopting green energy policies and improved waste management strategies is vital in promoting environmental sustainability within the sports industry. Utilizing renewable energy sources such as solar, wind, and geothermal power at sports venues aligns with global efforts to reduce carbon emissions and combat climate change. For example, using solar panels at stadiums and arenas contributes to energy efficiency while reducing dependence on non-renewable resources.

Additionally, effective waste management practices, such as recycling, composting, and adhering to the 3R Principle (Reduce, Recycle, Reuse), are essential for minimizing environmental harm. Significant events like the FIFA World Cup and the Olympics have increasingly focused on promoting waste minimization through compostable packaging, recycling-friendly signage, and reusable materials. Establishing comprehensive policies to integrate green energy solutions and responsible waste management can significantly enhance the industry's contribution to global sustainability goals. According to a research study, waste management in Thai Premier League football stadiums is ineffective, with less than 15% of recyclables processed while most are sent to landfills. Interviews revealed insufficient incentives, regulations, and awareness. The proposed framework highlights the need for policies, public participation, and resource recovery improvements (Journal of Population and Social Studies, 2017).

Sustainable Sports Events

As the world grapples with the pressing challenges of climate change, the sports industry has an opportunity to become a powerful driver of sustainability. Organizing sustainable sports events is one of the most impactful ways to reduce the sector's environmental footprint while setting an example for global audiences. Sports events can significantly mitigate their environmental impact by focusing on low-carbon, zero-waste practices and embracing renewable energy. This section explores practical strategies for organizing eco-friendly sports events and the importance of collaboration with local governments and communities. It highlights successful case studies of significant events that have successfully integrated sustainability. These efforts make sporting events more environmentally responsible and inspire change in other industries and communities worldwide.

COLLABORATION WITH LOCAL GOVERNMENTS AND COMMUNITIES

Achieving sustainability goals in sports events requires collaboration between organizations, local governments, and communities. These partnerships are essential to ensuring that environmental sustainability is woven into the event's planning and execution.

Importance of Local Partnerships in Achieving Sustainability Goals

- **Shared Responsibility:** Local governments support sustainability by providing the infrastructure and policy frameworks for green sports events. This may include facilitating the construction of sustainable transport systems, waste management programs, and energy-efficient technologies in sports venues.
- **Community Involvement:** Engaging local communities in sustainability creates a sense of shared responsibility. It helps communities feel more invested in the event's success, which can increase public support for environmentally friendly initiatives. For example, local

community groups might organize clean-up drives after events or participate in awareness campaigns related to sustainability.

Role of Sports Organizations in Influencing Policy and Infrastructure

Sports organizations can influence government policies and infrastructure by advocating for sustainability in their event planning. For instance, the *International Olympic Committee (IOC)* and *FIFA* have included sustainability criteria in the selection process for host cities (International Olympic Committee, 2020), requiring them to demonstrate their commitment to reducing environmental impacts. Additionally, these organizations can work closely with local governments to ensure that the necessary infrastructure, such as renewable energy sources and efficient transportation systems, is in place to support the event.

ATHLETES AND ORGANIZATIONS AS SUSTAINABILITY CHAMPIONS

The sports industry has a unique platform to advocate for change, with athletes and organizations wielding considerable influence over global audiences. Athletes, as admired public figures, are well-positioned to raise awareness about pressing issues such as climate change and environmental sustainability. They can inspire millions to take action toward a more sustainable future using their reach and popularity. Similarly, sports organizations and governing bodies increasingly recognize their role in promoting sustainability, setting important benchmarks for environmental responsibility through policies and initiatives that influence the broader industry. This section delves into how athletes and sports organizations can become champions for climate action and sustainability, examining their influence and efforts to drive meaningful change within and beyond the sports world.

Athletes as Advocates for Climate Action

Athletes have long been regarded as role models, and today, many are leveraging their public platforms to advocate for climate action. The power of an athlete's voice can inspire millions of fans to think more critically about their impact on the environment and encourage them to adopt more sustainable practices (Thomas & Hicks, 2024)

How Athletes Can Use Their Platforms to Raise Awareness About Climate Change and Promote Sustainability

- **Public Advocacy:** Athletes can use interviews, social media, and public appearances to speak out about the urgent need for climate action. With their ability to connect with diverse audiences, they can bring attention to key environmental issues, such as deforestation, pollution, and the impact of climate change on vulnerable communities. Their advocacy often extends beyond simple awareness, encouraging fans to participate in sustainable practices, such as reducing plastic use or supporting eco-friendly brands.

- **Influence on Policy and Action:** Many athletes have become involved in campaigns or initiatives to influence policy change. By partnering with environmental organizations or lobbying for sustainable practices in their respective sports, athletes can help drive large-scale efforts to tackle climate change. Their involvement often highlights the importance of government action to protect the environment.

Examples of Athletes Leading Sustainability Initiatives

Lewis Hamilton, a seven-time Formula One World Champion, has emerged as a passionate advocate for climate action. Beyond his racing achievements, Hamilton uses his global platform to raise awareness about the pressing issue of climate change. At events like WE Day UK in 2020, Hamilton urged young people to take collective responsibility for protecting the planet. He emphasized the importance of unity and making small, meaningful changes in daily life to combat the global climate crisis. Hamilton's journey into climate advocacy reflects a newfound purpose, where he acknowledges the privilege of living on this planet and the duty to treat it with care. His message is a powerful reminder of individuals' and communities' role in securing a sustainable future for generations.

Novak Djokovic is a world-class tennis champion and an advocate for sustainability and environmental consciousness. His transition to a plant-based lifestyle reflects his ethical commitment to addressing climate change. Djokovic emphasizes the significant role of animal agriculture in contributing to greenhouse gas emissions, deforestation, and environmental degradation (*Viva!*, 2021). By adopting a plant-based diet, he showcases how individual lifestyle choices can collectively contribute to reducing environmental harm. He believes in promoting awareness of the interconnectedness between our food systems, the welfare of animals, and the health of our planet. Djokovic's advocacy extends beyond performance reasons, underscoring a holistic approach to living responsibly and inspiring others to be more mindful of their ecological impact.

Tennis player Naomi Osaka promotes sustainable fashion by partnering with eco-friendly brands, while F1 driver Lewis Hamilton actively invests in renewable technologies and supports carbon offset initiatives (Thomas & Hicks, 2024).

CLIMATE ACTION AND ENVIRONMENTAL JUSTICE IN SPORTS

The role of sports organizations and athletes in promoting climate action goes beyond mere advocacy; it also involves addressing environmental justice issues. As influential figures with significant platforms, athletes can amplify voices from vulnerable communities disproportionately affected by climate change. Similarly, sports organizations can use their reach to highlight inequalities in access to climate-resilient infrastructure and promote inclusivity in sustainable development efforts.

Environmental justice in sports also ensures that eco-friendly practices are accessible to all, including developing regions where resources for sustainable initiatives may be limited. Initiatives promoting renewable energy adoption and waste management should aim for inclusivity, ensuring that local communities benefit from sustainability efforts rather than being excluded.

Collaborating with grassroots organizations, community leaders, and policymakers can enhance the effectiveness of climate action within sports. The sports industry can contribute to a more just and sustainable future by championing social equity alongside environmental responsibility.

Sports Organizations Leading by Example

In addition to athletes, sports organizations and governing bodies are increasingly taking steps to integrate sustainability into their operations. These organizations are setting key benchmarks that others in the industry can follow, and they often collaborate with governments and NGOs to scale their initiatives.

The FIFA World Cup is one of the most significant sporting events globally, and its environmental impact is substantial. For instance, the 2022 Qatar World Cup faced criticism for its carbon footprint, driven by international travel and the construction of air-conditioned stadiums (Sharma, 2023). However, FIFA has taken steps to integrate sustainability into its operations, such as carbon offset programs and promoting renewable energy use at venues (Cayolla & Escadas, 2023).

Similarly, the Olympic Games present both challenges and opportunities for sustainability. The Tokyo 2020 Olympics emphasized renewable energy, with venues powered by solar and hydrogen energy sources. Additionally, medals were crafted from recycled electronic waste, showcasing innovation in resource reuse (Geneva Environment Network, 2023). The upcoming Paris 2024 Olympics aims to set a benchmark for low-carbon operations, focusing on sustainable transportation, eco-friendly infrastructure, and minimizing waste (Francis et al., 2023).

CORPORATE SOCIAL RESPONSIBILITY (CSR) INITIATIVES IN THE SPORTS INDUSTRY

Sports organizations embrace corporate social responsibility (CSR) to contribute positively to environmental and social causes. CSR initiatives in the sports industry often focus on sustainability, from reducing a league's environmental footprint to supporting local communities through charitable projects.

- **The NFL's Environmental Initiatives:** The NFL PLAY 60 program, a collaboration between the American Heart Association and the National Football League, encourages youth to engage in 60 minutes of daily physical activity to improve overall health and wellness. Over 15 years, the initiative has reached millions through resources like the NFL PLAY 60 app, interactive digital experiences, and events like the Super Bowl Fitness Break. Educators can participate in challenges such as the Exercise Blitz, offering schools a chance to win grants for physical activity equipment. The program also emphasizes fitness education through on-demand content, featuring NFL teams demonstrating exercises, and integrates CPR awareness to promote lifesaving skills (*NFL PLAY 60*, 2020)
- **Formula 1's Sustainability Push:** F1 is also transitioning to sustainable fuels, planning 100% use by 2026, and improving logistics with biofuel trucks and sea freight. These efforts highlight F1's dedication to reducing its environmental impact while inspiring the global adoption of sustainable practices in sports and beyond. Aggreko's role aligns with its work in Formula E, further strengthening its leadership in energy transition solutions (Jessen, 2024)
- **The IOC's Sustainability Commitment:** The **Olympic Agenda 2020**, adopted by the IOC in December 2014, is a strategic roadmap comprising 40 recommendations aimed at safeguarding Olympic values, enhancing the role of sport in society, and making the Olympic Games future-ready. Built on the pillars of Credibility, Sustainability, and Youth, it was developed through a collaborative process involving stakeholders and experts. Since its adoption, the agenda has brought significant reforms, ensuring the Games are

adaptive, strengthening Olympic values, and expanding the societal impact of sport. Its implementation has laid a strong foundation for the continued evolution of the Olympic Movement (IOC, 2021)

- Athletes and sports organizations have the power to lead by example in advocating for climate action and sustainability. Athletes, with their global influence, can use their platforms to raise awareness and drive positive change. At the same time, sports organizations are setting important benchmarks by adopting sustainable practices across all areas of their operations, from event planning to corporate responsibility. As sustainability becomes an increasingly important issue worldwide, the sports industry's role as a champion for climate action will continue to grow, inspiring millions of fans and other industries to follow suit.

GRASSROOTS SPORTS AND COMMUNITY RESILIENCE

Grassroots sports initiatives have the potential to serve as powerful platforms for building climate-conscious communities and fostering long-term environmental resilience. These local programs often focus on engaging youth and community members in ways that encourage sustainable behavior while promoting the values of teamwork, discipline, and environmental stewardship. By integrating sustainability into community-based sports programs, it is possible to develop healthier, more active communities, raise awareness about climate change, and inspire collective action for a more sustainable future. This section explores how grassroots sports can be a catalyst for climate action through educational programs, community engagement, and creating long-term climate-resilient strategies.

Using Local Sports Initiatives to Build Climate-Conscious Communities

Local sports initiatives foster a deeper connection between individuals and their environment. When aligned with sustainability principles, these programs can be instrumental in educating communities about environmental issues and encouraging practices that reduce their carbon footprints.

How Local Sports Programs Can Help Educate and Engage Communities on Sustainability

The Sport for Unity, Resilience, and Empowerment (SURE) project is one of the finest examples. This project analyzes cooperation models that enhance resilience and inclusion, particularly in crisis settings. Partners from 13 countries, led by OPES (Italy) and 100% Sport (Austria), are researching best practices for inclusive sport and the lived experiences of vulnerable groups regarding access to sport. The project aims to identify ways to make grassroots sports more accessible and affordable and to prevent dropout during crises. Co-funded by the EU, it seeks to test a cooperation model to build resilience in the sports movement and ensure sports' societal role for vulnerable communities (Alessia De Iulis, 2024)

- **Integrating Sustainability into Sports Training**: Local sports clubs and recreational programs can incorporate environmental education into their coaching and training sessions. For example, children and youth in sports can be taught about sustainable practices such as reducing waste, conserving water, and protecting local ecosystems. Coaches can use the setting of sports to discuss climate change, pollution, and renewable energy, helping players understand the environmental challenges facing their communities. The Sustainable Training Methods for Athletes focuses on integrating eco-friendly practices into the sports industry to reduce environmental impact. This includes using recycled or biodegradable materials in sports gear, such as organic cotton and bamboo fibers, which offer comfort and durability while minimizing waste. Energy-efficient technologies like long-lasting wearables and innovative training equipment, including machine regenerative braking systems, help reduce electricity consumption. Likewise, virtual and augmented reality (VR/AR) tools have revolutionized training by offering immersive experiences that reduce the need for physical infrastructure and travel, thus lowering emissions. These technologies allow athletes to train effectively while contributing to sustainability efforts (Kazazi, 2024)

Adopting these sustainable methods in sports training reduces the carbon footprint and helps athletes reach their performance goals in an environmentally conscious way. Through eco-friendly gear, energy-efficient devices, and innovative training technologies, athletes can play a crucial role in promoting sustainability in sports.

- **Building Partnerships with Environmental Organizations:** Many grassroots sports programs collaborate with environmental organizations to create events and initiatives that combine sports and environmental stewardship. These partnerships can involve organizing cleanup activities, tree-planting drives, or recycling programs alongside sports events, creating a holistic approach to community sustainability (Yung, 2023)

COMMUNITY-BASED SPORTS PROGRAMS MAKING A DIFFERENCE

The **EFL Green Clubs**: The **EFL Green Clubs** initiative is a pioneering scheme supporting football clubs nationwide to enhance their environmental practices. The scheme encourages clubs to adopt sustainable practices in collaboration with GreenCode, an environmental accreditation program. This effort follows in the footsteps of Forest Green Rovers, recognized as the world's greenest football club by the United Nations and FIFA (EFL, n.d.). The scheme is designed to help clubs implement and improve environmental strategies, contributing to a greener future for football. Clubs that achieve GreenCode accreditation, such as Bristol City, Forest Green Rovers, and Sutton United, lead the way in adopting sustainable practices. The EFL is also actively improving its environmental performance, furthering its commitment to sustainability.

Through this program, more than 50 clubs, including prominent teams like Sheffield United, Portsmouth, and Millwall, are making significant strides toward reducing their environmental footprint. Even clubs that have left the EFL, such as Luton Town, remain committed to sustainability, retaining their membership in the Green Clubs initiative for the 2023/24 season.

- **World Rugby:** World Rugby is deeply committed to addressing environmental sustainability through its Environmental Sustainability Plan 2030. The plan outlines concrete actions that rugby organizations globally take to minimize environmental impact. Given its global reach and strong community presence in over 120 countries, the sport is uniquely positioned to contribute to climate change action.

The Environmental Sustainability Plan 2030 serves as a statement of intent and a roadmap to tackle environmental challenges. It is based on three key priorities: Climate Action, Circular Economy, and Natural Environment Protection. World Rugby emphasizes addressing carbon footprints, advocating for climate awareness, reducing waste, and promoting ecosystem sustainability.

Notably, the plan is committed to addressing direct impacts, delivering sustainable rugby events, and educating and advocating for sustainability throughout rugby's global community. The strategy ensures that rugby contributes positively to a healthier, more sustainable society while promoting solidarity and environmental respect.

Through partnerships with organizations like ChildFund Rugby and initiatives like Rugby for Nature, World Rugby is working to involve the entire rugby family in climate and environmental efforts. The Rugby World Cup and U20 Championship are also platforms to spread awareness and engage in actions that have a positive environmental impact.

This comprehensive approach to sustainability reflects rugby's dedication to making a difference beyond the game and actively participating in the global effort to combat climate change (worldrugby.org, n.d.)

Sports as Vehicles for Climate Action Education

Sports programs provide an ideal platform to educate young people about climate change and its implications, as they combine physical activity with educational content. These programs can instil a sense of environmental responsibility while simultaneously promoting health and wellness.

The World Olympians Association (WOA) is spearheading a significant global initiative, #Sport4Climate, aimed at raising awareness about climate change and leveraging the influence of athletes to drive environmental change. The campaign underscores the powerful role that sports can play in highlighting environmental issues, particularly through the widespread reach of social media and the status of Olympic athletes as influential public figures.

Leszek Jan Sibilski, a former member of Poland's national cycling team, is one of the prominent advocates for the campaign. He points out the visible impacts of global warming on the world of sports, referencing instances like the melting snow at the Sochi 2014 Winter Olympics and heat exhaustion among players during the 2014 Australian Open Tennis Championships as early signs of climate change's effects on sporting events.

Sibilski emphasizes that Olympians, due to their high-profile status, can serve as powerful role models and advocates for change. Their involvement in the #Sport4Climate initiative encourages them to utilize their social media platforms to spread the message and inspire action against climate change. Notable athletes, including Thiago Pereira, a Brazilian Olympian, have already voiced their support for the campaign, contributing to its momentum.

The initiative includes an 'Athletes' Pledge' for Olympians, who are encouraged to commit to promoting sustainability and environmental responsibility both within their practices and through their public influence. This reflects the broader philosophy of the WOA, which seeks to engage Olympians in global social issues, particularly through platforms that amplify their voice and leadership.

The WOA continues to work on initiatives that address other global challenges, including the promotion of sports diplomacy and gender-based violence (GBV). The WOA also offers scholarships and organizes the OLY House Film Festival to highlight impactful stories related to sports and diplomacy.

By fostering a collective effort among athletes, the #Sport4Climate campaign aligns with the Olympic values of solidarity, respect, and fair play while emphasizing the urgency of tackling climate change on a global scale.

POLICY FRAMEWORKS AND GLOBAL STANDARDS FOR SUSTAINABLE SPORTS

As the sports industry continues to evolve, there is a growing recognition of the need for structured, global policies and frameworks to support sustainability. These frameworks not only provide guidance but also set measurable standards that can drive change within the sports sector. From international certifications like ISO and LEED to governmental regulations, policies at both the national and international levels are increasingly pushing for sustainable practices in sports events, facilities, and operations (4C Consulting, 2024). This section will explore how these frameworks are shaping the future of sustainable sports, discuss the role of policymakers in advancing green sports practices, and examine the impact of global standards in fostering sustainability across the industry.

International Sustainability Frameworks and Certifications

The adoption of internationally recognized sustainability frameworks and certifications plays a critical role in promoting environmental responsibility within the sports sector. These frameworks offer tangible metrics for organizations to measure their environmental impact and set targets for improvement. By aligning with such standards, sports organizations can improve their environmental performance, earn credibility, and inspire others to follow suit.

Overview of Existing Global Standards for Sustainability in Sports:

Sustainability certifications play a vital role in demonstrating a company's commitment to environmental responsibility and social equity. In today's world, where both consumers and stakeholders demand greater transparency, these certifications provide a credible way for businesses to prove their efforts toward sustainability. By adhering to recognized standards, companies not only comply with regulatory requirements but also position themselves as leaders in responsible business practices. These certifications cover a range of areas, from environmental protection and energy efficiency to fair labour practices, and help businesses showcase their contributions to a sustainable future.

a. ISO 14001 (Environmental Management System)

Overview: ISO 14001 is one of the most widely adopted certifications for environmental management systems (EMS). Issued by the International Organization for Standardization (ISO), it helps organizations identify, manage, and reduce their environmental impacts. The certification encourages continual improvement in environmental performance focusing on Legal compliance, waste reduction, resource efficiency, and minimizing environmental impact. Iso has a global reach of over 360,000 organizations worldwide hold this certification. ISO helps businesses ensure they meet legal requirements, manage risks, and enhance their environmental impact in a structured manner. It also demonstrates a company's commitment to sustainability to both customers and regulators.

b. Leadership in Energy and Environmental Design (LEED)

LEED is a certification from the U.S. Green Building Council (USGBC) aimed at promoting sustainability in the construction and building sector. It provides a framework for the design, construction, and operation of green buildings, focusing on energy efficiency, resource conservation, and a reduction in environmental impact.it focuses on sourcing of sustainable materials, energy and water efficiency, indoor environmental quality, and site sustainability. LEED is internationally recognized and applied in over 160 countries. LEED-certified buildings are proven to be more energy-efficient, reducing operational costs for owners. Additionally, they attract tenants who value environmental sustainability.

c. ENERGY STAR

ENERGY STAR is a certification that signifies a building's energy efficiency, meaning it performs better than 75% of similar buildings in the country in terms of energy consumption. This certification is based on a performance score given by the Environmental Protection Agency (EPA). ENERGY STAR mainly focuses on energy efficiency, greenhouse gas reduction, and sustainability in building operations, primarily used in the United States, but its influence is growing globally. Buildings with ENERGY STAR certification are more cost-effective to operate and are more attractive to tenants looking to lower their carbon footprint.

d. B Corporation (B Corp)

B Corporation is a certification given to companies that meet high standards of social and environmental performance, accountability, and transparency. This certification is awarded by the nonprofit B Lab to for-profit companies that demonstrate a strong commitment to sustainability and social responsibility. B Corp focuses on environmental performance, social responsibility, ethical practices, and corporate governance. It has a reach of over 8,100 companies that are B Corp certified across 95 countries. B Corp certification indicates a company's dedication to holistic sustainability, promoting ethical practices that benefit society, workers, and the environment. It also helps businesses attract consumers and investors who are interested in sustainability.

e. FairTrade

FairTrade certification ensures that the production of goods such as coffee, chocolate, and cotton is done ethically, with fair wages for workers, sustainable farming practices, and a focus on improving the living standards of small farmers and businesses. It focuses on the aspects of fair wages, community development, environmental protection, and improved working conditions in developing countries. FairTrade certification is recognized globally and applied to thousands of consumer goods. It supports the creation of equitable global supply chains and guarantees fair compensation and working conditions for marginalized producers. For consumers, it offers the assurance that their purchases contribute to global social justice.

f. Rainforest Alliance Certified™ Seal

The Rainforest Alliance certification focuses on promoting sustainable agricultural, forestry, and tourism practices. It emphasizes environmental conservation, social equity, and economic viability, ensuring that businesses protect ecosystems and biodiversity while promoting social fairness and improving local economies. The Seal having reached over 70 countries across the world focuses on the area of environmental protection, social fairness, and economic sustainability, including wildlife preservation, reducing pesticide use, and promoting sustainable farming practices. It promotes sustainable land use practices and supports communities in maintaining livelihoods while preserving their local ecosystems.

g. Forest Stewardship Council (FSC)

The FSC is an international nonprofit organization focused on promoting sustainable forestry practices worldwide. The certification ensures that wood and forest products come from responsibly managed forests that provide environmental, social, and economic benefits. It mainly focuses on the areas of forest conservation, biodiversity protection, workers' rights, and sustainable forest management. The FSC operates in more than 80 countries and covers various sectors, including forestry, paper production, and furniture manufacturing. FSC ensures that consumers can confidently choose wood products that have been responsibly sourced, knowing that they have been produced in a manner that supports environmental sustainability, fair labour, and local communities.

h. Green Seal

Green Seal is a certification awarded by a U.S.-based nonprofit organization that verifies the environmental sustainability of products and services. It covers a wide range of industries, including cleaning products, paints, and building materials. While it is U.S.-focused, Green Seal is influential in many regions globally mainly in waste reduction, energy efficiency, pollution reduction, and promoting health safety. This certification helps consumers identify products and services that meet rigorous environmental and health criteria, ensuring their choices contribute to a cleaner, healthier environment.

i. SA8000

SA8000 is a certification standard focused on ensuring ethical practices in the workplace. It covers a wide range of social accountability issues, including child labour, forced labour, health and safety, discrimination, and working hours. SA8000 used across various countries focuses on workers' rights, fair wages, safe working conditions, and human rights protections. Companies with this certification are recognized for treating their employees fairly, ensuring their rights are upheld, and contributing to the well-being of their workers. It builds consumer confidence and promotes ethical sourcing.

Each of these certifications highlights the growing demand for transparency and accountability in corporate sustainability. They cover various aspects of business, from environmental impact and energy efficiency to social fairness and worker welfare. Achieving these certifications helps businesses build credibility, enhance their reputation, and meet the expectations of increasingly eco-conscious consumers and investors. As the world continues to face global challenges like climate change, these certifications serve as vital tools for companies striving to make a positive environmental and social impact.

Shaping Industry Practices and Setting Global Standards

These global standards are playing a pivotal role in influencing the sports industry to adopt sustainable practices. They not only guide organizations in reducing their environmental impact but also provide a benchmark for comparing performance across the industry. By meeting these standards, sports organizations enhance their reputation, attract eco-conscious sponsors, and ensure they are aligned with global sustainability goals. For instance, the International Olympic Committee (IOC) has integrated sustainable development into its planning process, requiring all Olympic bid cities to meet specific environmental criteria.

The Role of Policymakers in Advancing Sustainable Sports Practices

While sustainability frameworks and certifications provide practical guidelines, policymakers' involvement is essential for creating an enabling environment for green sports practices. Governments at various levels can incentivize sustainable behaviors, set regulatory requirements, and allocate resources for sustainability initiatives in sports.

- **The Importance of Policy Intervention in Encouraging Sustainability in Sports:** Policy intervention is critical to driving large-scale sustainability across the sports sector. Governments can introduce regulations and incentives encouraging sports organizations to adopt eco-friendly practices. These policies can include mandates for energy-efficient buildings, waste reduction, and carbon footprint reporting for large-scale events and sports facilities. Furthermore, policymakers can foster collaboration between the public and private sectors to fund and support sustainability initiatives in sports.
- **National and International Policies Promoting Green Sports Practices:** The European Green Deal, announced in 2019, outlines the EU's strategy to achieve climate neutrality by 2050 by cutting greenhouse gas emissions by 55% by 2030 compared to 1990 levels (European Parliament, 2022). The "Fit for 55" package, with 13 interlinked policies, aims to overhaul sectors like energy, transport, and industry. It includes reforms to the Emissions Trading System, carbon border adjustments, and a commitment to increase renewable energy consumption to 42.5% by 2030. The EU plans to mobilize over €1 trillion for these efforts, promoting a circular economy and preserving biodiversity.

- The Paris Agreement for Sport and Sustainable Development, introduced before the Paris 2024 Olympic Games, aims to leverage sports to support the UN's Sustainable Development Goals by 2030. Launched by French President Emmanuel Macron and IOC President Thomas Bach, it brought together 500 participants, including heads of state, international organization leaders, and athletes (Lavier, 2024). The agreement establishes ten commitments focused on five key priorities, intending to integrate sustainability into sports globally.
- **Education and Employment**: Promoting the use of sport to provide quality education and employment opportunities to youth globally.
- **Health and Nutrition**: Encouraging sports as a medium to enhance global health outcomes and improve nutrition, particularly for marginalized communities.
- **Equality and Inclusion**: Ensuring that sport is a vehicle for promoting gender equality and social inclusion and ensuring accessibility for all, regardless of background or ability.
- **Financing and Impact Measurement**: Establishing financial commitments to support sustainable initiatives within sport, as well as systems for measuring the long-term impact of these initiatives.
- **Sustainability and Legacy**: Integrating environmental sustainability into the core of sporting practices, aiming to create lasting positive legacies through responsible resource use and ecological responsibility.

The agreement emphasizes that sport, a universal language that unites diverse communities, must contribute to entertainment and serve as a force for positive change. It aims to accelerate progress in all areas of sustainable development by fostering a global collaboration between governments, civil society, and international sporting bodies.

Central to the agreement's vision is ensuring that sport not only meets its stakeholders' immediate needs but also safeguards future generations' potential. This requires a holistic approach that balances environmental, social, and economic factors to create a prosperous, healthy, and equitable society. Through the agreement, sports organizations are encouraged to align their actions with the SDGs, thus ensuring that sport is accessible to everyone, everywhere, while contributing to building a better, more sustainable world.

The Paris Agreement also provides an important framework for collaboration within the Olympic Movement, fostering deeper partnerships and shared commitments between sports federations, governments, the private sector, and civil society. This collective action is critical for creating a legacy of sustainability that ensures sports continue to play a transformative role in society long into the future.

International sustainability frameworks and certifications, combined with government-supportive policies, are key drivers of sustainability in the sports sector. These frameworks provide guidelines for sports organizations to reduce their environmental impact and set global standards that influence industry practices. Policymakers are critical in advancing sustainable sports practices by enacting regulations, providing incentives, and fostering collaboration. As more countries and organizations align with global sustainability goals, the sports industry can become a powerful force for climate action and environmental responsibility.

CONCLUSION

This chapter has explored the transformative potential of the sports industry in driving global sustainability efforts. From examining the environmental footprint of sports to highlighting innovative practices in sustainable infrastructure, event management, and grassroots initiatives, it is clear that the sports sector holds significant promise in fostering a greener and more sustainable future. The key takeaways from this chapter are:

- **Sustainability is Essential:** The sports industry, like many others, faces challenges in reducing carbon emissions, waste, and resource consumption. However, there are immense opportunities to reduce its environmental impact by embracing renewable energy, sustainable materials, and innovative waste management practices.
- **Collaboration is Key:** The role of athletes, sports organizations, local communities, and policymakers is crucial in driving sustainability. Whether advocating for climate action, adopting green building

standards, or developing policies that support eco-friendly practices, every stakeholder has a part to play.

- **Innovative Solutions Are Emerging:** From eco-friendly stadiums and low-carbon events to community-driven sports programs, there are numerous examples of how sports organizations implement sustainable solutions. These innovations demonstrate that sustainability in sports is not just a possibility but a growing reality.
- **Global Standards and Policies Are Shaping the Future:** International sustainability frameworks and government policies provide the guidance and resources to push the sports sector towards a more sustainable future. The industry is held to higher environmental standards through certification programs and national regulations, ensuring sustainability is embedded in sports culture worldwide.

Alignment with Sustainable Development Goals (SDGs)

The sports industry's ongoing efforts toward sustainability directly contribute to achieving multiple Sustainable Development Goals (SDGs) established by the United Nations. Notably, the industry's focus on reducing carbon emissions, adopting green energy solutions, and minimizing waste generation aligns well with:

- **SDG 13 (Climate Action):** By implementing renewable energy practices, promoting eco-friendly travel, and advocating for carbon-neutral events, the sports industry contributes to mitigating climate change and enhancing resilience to climate-related impacts.
- **SDG 12 (Responsible Consumption and Production):** The shift toward sustainable event management, including recycling initiatives, compostable packaging, and energy-efficient infrastructure, supports the goal of promoting responsible consumption and minimizing environmental harm.
- **SDG 11 (Sustainable Cities and Communities):** Collaborative efforts between local governments, sports organizations, and communities to create sustainable sports facilities and transpor-

tation networks contribute to building resilient, sustainable urban environments.

- **SDG 7 (Affordable and Clean Energy):** By incorporating clean energy technologies, such as solar and wind power, sports facilities can reduce their reliance on non-renewable resources and promote energy efficiency.

Integrating these SDGs into sports management practices addresses environmental concerns and is a powerful tool for raising awareness and inspiring global audiences to adopt more sustainable lifestyles. The sports industry's unique ability to influence cultural norms and behavior makes it a critical partner in achieving global sustainability targets.

CALL TO ACTION

For sports managers, policymakers, athletes, and all stakeholders in the sports industry, the time to act is now. By working together, we can harness the power of sports to reduce its environmental footprint and inspire global action on climate change. Sports managers can implement green practices in facility management and event planning. Policymakers must continue to strengthen the regulatory framework for sustainable sports. Athletes, as influential public figures, have the power to raise awareness and lead by example. Through collective effort, we can make the sports industry a leader in sustainability and a powerful force for positive environmental change.

POLICY RECOMMENDATIONS FOR SUSTAINABLE SPORTS PRACTICES

Establishing well-defined policies and frameworks is essential to promote sustainability within the sports industry. These policies should address energy consumption, carbon emissions, waste management, and stakeholder collaboration. The following recommendations can guide sports organizations, policymakers, and event organizers toward achieving sustainability goals:

Adopting Renewable Energy Standards

The sports industry must transition toward cleaner energy sources to reduce its carbon footprint. Policies should encourage the adoption of renewable energy technologies, such as solar panels, wind turbines, and geothermal systems, especially in large stadiums and sports complexes.

- **Incentive Mechanisms:** Governments can offer tax breaks, subsidies, or grants to sports organizations that demonstrate substantial progress in reducing their carbon emissions by integrating renewable energy systems.
- **Public-Private Partnerships:** Collaborations between sports organizations, energy providers, and environmental agencies can accelerate the implementation of green energy projects.
- **Monitoring and Reporting:** Establishing mandatory reporting standards for carbon emissions and energy usage will promote transparency and accountability, encouraging sports entities to work towards sustainability actively.

Sustainable Waste Management Protocols:

Effective waste management is critical to minimizing the environmental impact of sports events. Policies should enforce proper disposal, recycling, and reuse of materials used during events.

- **Zero-Waste Guidelines:** Major sports events should be required to submit sustainability plans that include strategies for achieving zero-waste operations.
- **Biodegradable and Recyclable Materials:** Using eco-friendly materials for packaging, merchandising, and promotional activities can significantly reduce plastic waste.
- **Certification and Recognition:** Developing certification schemes to recognize sports events that successfully implement sustainable

waste management practices will motivate other organizers to follow suit.

- **Waste Management Education:** Public awareness campaigns involving athletes and sports organizations can promote responsible waste disposal and encourage recycling practices among fans and local communities.

Climate-Resilient Infrastructure Development:

The increasing frequency of extreme weather events due to climate change makes it necessary for sports infrastructure to be resilient and sustainable. Policies must address construction standards, maintenance practices, and environmental considerations.

- **Eco-Friendly Construction Materials:** Promoting the use of sustainable materials like bamboo, recycled steel, and low-carbon cement in the construction of sports facilities can reduce the industry's carbon footprint.
- **Climate-Adaptive Designs:** Encouraging the integration of designs that can withstand natural disasters, such as flooding, heatwaves, and heavy rainfall, can enhance the resilience of sports infrastructure.
- **Funding and Technical Assistance:** Governments should provide financial incentives and technical support to local sports facilities aiming to upgrade their infrastructure to meet sustainability standards.
- **Green Building Certifications:** Encouraging sports facilities to obtain certifications like LEED (Leadership in Energy and Environmental Design) or ISO 14001 for their sustainability efforts will enhance credibility and promote broader adoption of eco-friendly practices.

Inclusive Sustainability Policies

Sustainability initiatives must be accessible to all, including underprivileged communities and developing regions. Policies should promote inclusivity by ensuring that green practices benefit local populations rather than marginalize them.

- **Accessibility and Affordability:** Ensuring that renewable energy solutions and eco-friendly infrastructure are available to grassroots sports organizations and community facilities.
- **Equitable Resource Allocation:** Governments and sports organizations must collaborate to provide funding and resources for developing sustainable sports infrastructure in underserved areas.
- **Community Engagement Programs:** These programs encourage local participation in green initiatives through education, awareness campaigns, and accessible training programs.
- **Environmental Justice Frameworks:** Recognizing the disproportionate impact of climate change on marginalized communities and ensuring that sustainability efforts do not contribute to social or economic inequality.
- **Collaboration and Stakeholder Engagement:** Effective policies require cooperation between various stakeholders, including sports organizations, governments, private entities, and communities.
- **Public-Private Partnerships:** Collaborative efforts between sports organizations, local governments, environmental agencies, and corporate sponsors can facilitate resource-sharing and innovation in sustainable practices.
- **Athlete Advocacy:** Athletes can play a crucial role in promoting environmental responsibility by supporting public campaigns, endorsing eco-friendly brands, and encouraging their followers to adopt sustainable practices.
- **Knowledge Exchange Platforms:** Establishing forums where experts, policymakers, and sports professionals can share insights and best practices for implementing sustainable solutions in sports.
- **Collaborative Research:** Encouraging research collaborations between academic institutions, environmental organizations, and sports bodies to develop innovative methods for achieving climate goals.
- Implementing these policy recommendations can significantly enhance the sports industry's capacity to contribute to global sus-

tainability efforts. Establishing clear guidelines, incentives, and collaborative frameworks will ensure that sports continue to play a transformative role in achieving climate and environmental goals.

Future Outlook

Looking ahead, the potential for sports to serve as a central platform for global climate advocacy and sustainable development is immense. As global attention shifts toward sustainability, sports organizations have the unique opportunity to engage millions of fans, inspire communities, and drive meaningful change. The future of sports is one where environmental consciousness is embedded in every aspect of the industry—from the construction of facilities to how events are managed and how athletes advocate for sustainability. By fully embracing these challenges and opportunities, sports can catalyze a more sustainable, resilient, and climate-conscious world. The journey has already begun, and the sports industry is poised to create a greener, more sustainable future for generations to come.

ACKNOWLEDGEMENT

This content has been created without the use of generative AI. All ideas and text are original and authored by the contributors, with proper citations provided where applicable. Grammarly has assisted with language enhancement and ensured clarity in English writing.

REFERENCES

4C Consulting. (2024, September 26). *ISO 50001 Certification: Boosting Green Building Efficiency for LEED Certification | 4C Consulting*. 4C Consulting. https://www.4cpl.com/blog/iso-50001-certification-boosting-green-building-efficiency-for-leed-certification/

. Abbass, K., Qasim, M. Z., Song, H., Murshed, M., Mahmood, H., & Younis, I. (2022). A Review of the Global Climate Change impacts, adaptation, and Sustainable Mitigation Measures. *Environmental Science and Pollution Research, 29*(1), 42539–42559. Springer. DOI: 10.1007/s11356-022-19718-6

Alessia De Iulis. (2024, October 8). *The European Non-Governmental Sports Organisation (ENGSO)*. ENGSO | the European Organisation for Grassroots Sport. https://engso.eu/sport-for-unity-resilience-and-empowerment-will-analyse-cooperation-models-focusing-on-resilience-and-inclusion/

. Cayolla, R., & Escadas, M. (2023). Environmental Sustainability and Sports Management: A Review of Marketing Contributions and Discussion of Future Research Opportunities. *Smart Innovation, Systems and Technologies*, 309–321. DOI: 10.1007/978-981-99-0333-7_23

EFL. (n.d.). *EFL Green Clubs*. EFL. https://www.efl.com/about-the-efl/green-clubs/

European Parliament. (2022, June 22). *Green Deal: key to a climate-neutral and sustainable EU*. European Parliament. https://www.europarl.europa.eu/topics/en/article/20200618STO81513/green-deal-key-to-a-climate-neutral-and-sustainable-eu

FIFA. (2022). *Sustainability: Waste management at tournament sites*. FIFA Publications. https://publications.fifa.com/en/final-sustainability-report/environmental-pillar/waste-minimisation/waste-management-at-tournament-sites/

Francis, A. E., Webb, M., Desha, C., Rundle-Thiele, S., & Caldera, S. (2023). Environmental Sustainability in Stadium Design and Construction: A Systematic Literature Review. *Sustainability (Basel)*, *15*(8), 6896. DOI: 10.3390/su15086896

Gamage, A., Gangahagedara, R., Subasinghe, S., Gamage, J., Guruge, C., Senaratne, S., Randika, T., Rathnayake, C., Hameed, Z., Madhujith, T., & Merah, O. (2024). Advancing sustainability: The impact of emerging technologies in agriculture. *Current Plant Biology*, *40*, 100420. DOI: 10.1016/j.cpb.2024.100420

Geneva Environment Network. (2023, June 1). *Sports, the Environment, and the Role of International Geneva*. Www.genevaenvironmentnetwork.org. https://www.genevaenvironmentnetwork.org/resources/updates/sports-the-environment-and-the-role-of-international-geneva/

HOW THE 2020 TOKYO OLYMPICS SUSTAINABILITY AND ENVIRONMENTAL AWARENESS. (2020). Gamers. https://vocal.media/gamers/how-the-2020-tokyo-olympics-sustainability-and-environmental-awareness

Inside FIFA. (2023). Fifa.com. https://inside.fifa.com/social-impact/sustainability/fwc-2026-strategy/strategy

International Energy Agency. (2024, August 2). *Greenhouse Gas Emissions from Energy Data Explorer – Data Tools*. IEA. https://www.iea.org/data-and-statistics/data-tools/greenhouse-gas-emissions-from-energy-data-explorer

International Environmental Technology. (2018). *What Recycling Measures Are in Place at the World Cup?* Envirotech Online; Environmental. https://www.envirotech-online.com/news/business-news/44/breaking-news/what-recycling-measures-are-in-place-at-the-world-cup/46570

International Olympic Committee. (2020). *Requirements to host the Olympic Games*. Olympics.com. https://olympics.com/ioc/becoming-an-olympic-games-host/requirements-to-host-the-olympic-games

IOC. (2021, May 17). *Olympic Agenda 2020 - Strategic Roadmap for the Olympic Movement*. International Olympic Committee. https://olympics.com/ioc/olympic-agenda-2020

Jessen, J. (2024, August 29). *F1 Accelerates Sustainability in Partnership with Aggreko*. Sustainabilitymag.com; Bizclik Media Ltd. https://sustainabilitymag.com/articles/f1-accelerates-sustainability-in-partnership-with-aggreko

Journal of Population and Social Studies. (2017). *Sustainable Solid Waste Management in Sports Events: A Case Study of Football Matches in Thailand.* Retrieved from https://so03.tci-thaijo.org/index.php/jpss/article/view/102290

Kalair, A. R., Seyedmahmoudian, M., Stojcevski, A., Abas, N., & Khan, N. (2021). Waste to energy conversion for a sustainable future. *Heliyon*, *7*(10), e08155. DOI: 10.1016/j.heliyon.2021.e08155 PMID: 34729426

Kazazi, M. (2024, December 18). *Sustainable Training Methods for Athletes*. The Sports Financial Literacy Academy. https://moneysmartathlete.com/athletes-sustainable-practices-in-sports/sustainable-training-methods-for-athletes/

Khalid, W., & Rogstam, J. (2013). Energy usage prediction model comparing outdoor vs. indoor ice rinks. *Energy and Building*, *67*, 195–200. DOI: 10.1016/j.enbuild.2013.08.004

Lavier, T. (2024, July 25). *ITTF Welcomes Paris Agreement for Sport and Sustainable Development - International Table Tennis Federation.* International Table Tennis Federation. https://www.ittf.com/2024/07/25/ittf-welcomes-paris-agreement-for-sport-and-sustainable-development/

Lokesh Bohra. (2024, December 27). *Sustainable Play: Navigating ESG Trends in the Sports Industry*. Medium. https://medium.com/@lokeshbohra/sustainable-play-navigating-esg-trends-in-the-sports-industry-december-2024-9cbb7490fdf6

Mercedes-Benz Stadium. (2024). *Sustainability at Mercedes-Benz Stadium | Leaders in Zero Waste*. Www.mercedesbenzstadium.com. https://www.mercedesbenzstadium.com/sustainability

Naylor, C. (2024, August 7). Sports Event Food Waste: An Olympic-Sized Problem | BioteCH4. *Biotech, 4*, •••. https://biotech4.co.uk/sports-event-food-waste/

NFL PLAY 60. (2020). Www.heart.org. https://www.heart.org/en/professional/educator/nfl-play-60

Olympians' support for #Sport4Climate campaign gathers pace. (2025). Olympians.org. https://olympians.org/news/183/olympians-support-for-sport4climate-campaign-gathers-pace/

Olympics. (2024). *Carbon Neutral Games - Paris 2024*. Olympics.com. https://olympics.com/en/paris-2024/our-commitments/the-environment/carbon-method

Schaal, T. (2024, February 9). *Environmental Impact of the Super Bowl - OWL ESG*. OWL ESG. https://owlesg.com/2024/02/09/environmental-impact-of-the-super-bowl/

Sharma, S. (2023, February 17). *What is the Carbon Footprint of Sport? - The Carbon Literacy Project*. The Carbon Literacy Project - Relevant Climate Change Learning for Everyone. https://carbonliteracy.com/what-is-the-carbon-footprint-of-sport/

Sustainability» About the SDG's. (n.d.). https://sustainability.sport/about-the-sdgs

Sustainable Travel International. (2024). *Carbon footprint of tourism*. Sustainable Travel International. https://sustainabletravel.org/issues/carbon-footprint-tourism/

This Earth Day, Lewis Hamilton is fighting climate change. (n.d.). Https:// Www.we.org. https://www.we.org/en-US/we-stories/we-day/this-earth-day-lewis-hamilton-is-fighting-climate-change

Thomas, L., & Hicks, W. (2024, December 10). *Can sport empower us to be more sustainable?* World Economic Forum. https://www.weforum.org/stories/2024/12/can-sport-empower-us-to-be-more-sustainable/

Viva! (2021). Viva! Novak Djokovic. https://viva.org.uk/lifestyle/living-vegan/celebrities/sportspeople/novak-djokovic/

Wilby, R. L., Orr, M., Depledge, D., Giulianotti, R., Havenith, G., Kenyon, J. A., Matthews, T. K. R., Mears, S. A., Mullan, D. J., & Taylor, L. (2022). The impacts of sport emissions on climate: Measurement, mitigation, and making a difference. *Annals of the New York Academy of Sciences*, *1519*(1), 20–33. DOI: 10.1111/nyas.14925 PMID: 36377356

worldrugby.org. (n.d.). *Environmental Sustainability Plan 2030 | World Rugby*. Www.world.rugby. https://www.world.rugby/organisation/sustainability/environment

Yung, J. (2023, April 12). *Sport can support sustainability, contribute to environment-related SDGs*. Sportanddev. https://www.sportanddev.org/latest/news/sport-can-support-sustainability-contribute-environment-related-sdgs

Chapter 3
A Brief Study on Waste Management and Circular Policies in the Tourism Industry

Shivam Bhartiya
https://orcid.org/0000-0001-5532-2283
Jain University, India

Seda Yıldırım
https://orcid.org/0000-0003-4367-6652
Tekirdag Namık Kemal University, Turkey

Kushal Singh
Tripura University, India

ABSTRACT

The tourism industry, which is a significant contributor to the acceleration of climate change and the increase in environmental pollution, must also shoulder some responsibility for mitigating this. The tourism industry is considered a comprehensive and diverse industry encompassing a wide range of activities and services. Various activities, such as transportation, dining, accommodation, and travel, as well as tourist behavior, necessitate a comprehensive examination of the tourism industry's green trans-

DOI: 10.4018/979-8-3373-9685-9.ch003

formation. This study aims to conduct a specific analysis of the tourism industry regarding waste management and waste management policies. In this study, important waste management application examples and recycling strategies will be reviewed for the tourism industry worldwide. By categorizing the main areas within the tourism industry, specific waste management strategies and practices for these subcategories and areas will be discussed. The study focuses on establishing a basic profile for waste management and recycling policies in the tourism industry.

INTRODUCTION

Waste management policies, the focal point of the circular economy, have penetrated all sectors, making sustainable development goals realistic. Waste management policies and recycling practices, as circular practices, have become fundamental environmental and sustainability practices across industries and sectors (Islam et al., 2024; OECD, 2019; Laureti et al., 2024; Yıldırım, 2025; Munonye, 2025). In other words, businesses pursuing a circular economy have redesigned their production methods in line with circular practices. The tourism sector, a major contributor to environmental pollution (Bange et al., 2022; Yıldırım et al., 2023), stands out as one of the key sectors being redesigned in line with the circular economy.

Carbon emissions, solid waste, marine pollution, air pollution, water consumption, food waste, and post-production and post-consumption waste resulting from other functions and sub-service categories necessitate the tourism sector to adopt a circular economy model (Yıldırım et al., 2021; Rodríguez et al., 2020; Ahmad, & Ma, 2021). Studies conducted over the last decade have proven that the tourism sector is a serious environmental polluter. For example: Katircioglu (2014) conducted an econometric study demonstrating the link between tourism and environmental pollution in the context of the Turkish case. Empirical data suggest that tourism development increases energy use and has amplifying effects on climate change. Zhang and Gao (2016) examined the impacts of tourism-driven growth on energy and the environment in the context of the Chinese case study. The study suggested that the tourism sector, which is seen as essential for carbon emissions and economic growth, should be redesigned according to long-term sustainable policies. Growth driven by low-carbon tourism could be more sustainable in the long term. Metilelu & Michael (2022) reached some findings regarding tourism and pollution based on the Nigerian case study. The study stated that environmental pollution should also be considered when supporting coastal tourism. Unplanned excessive tourism activity causes pollution and environmental degradation on the coasts. Azam et al. (2018) conducted a study investigating the impact of tourism-related growth on environmental pollution, conducting econometric analysis for Malaysia, Singapore, and Thailand. The study concluded that tourism-related environmental pollution in Malaysia increased, while tourism-related environmental pollution was absent in Thailand and Singapore. Accordingly, following sustainable practices fosters growth through tourism while also protecting the environment (Azam et al., 2018). Gao & Zhang, (2021) conducted a study on content that causes tourism and air pollution for countries in the Mediterranean region. Valid results have been found for some Mediterranean countries where air polluting factors and tourism activities are related.

On the other hand, current literature contains many studies on the relationships between waste management, circular economy and the tourism sector. Chiwaridzo and Han (2025) studied sustainable practices in the tourism sector for Zimbabwe. They conducted a survey and they determined sustainable practices including policies for reducing carbon emission, waste and energy consumption could contribute to the tourism sector positively in the long term. Abdulkadir et al. (2025) studied some variables related to waste management in the tourism sector in Zanzibar. They also made recommendations for implementing sustainable waste management policies in the hotel sector. Expanding waste management practices in the tourism sector is crucial for sustainability. The tourism sector should address the shortage of trained personnel and infrastructure required for waste management and circular practices, and support the green transformation (Zorpos et al., 2021). Rodriguez et al. (2020) mentioned the key role of circular economy model in tourism sector. Camilleri (2021) examined food loss, waste management, and sustainability conditions in the tourism area and determined the importance of the green transition in the tourism sector. Strippoli et al. (2024) gave evidences for the need of circular economy model adaptation in tourism sector as considering the negative outcomes of tourism sector for the environment. Giurea et al. (2022) gave some evidences for the link between waste, energy and agro-tourism issues in the context of circular economy.

Aim and Design

New theoretical studies have begun to emerge in the tourism field, or in the tourism industry, to adapt to the circular economy. Research on the application of waste management practices, the focus of the circular economic model, to the tourism sector and its sub-sectors has been significant (Axhami et al., 2023; Falcone, 2019; Xu et al., 2022; Sørensen et al., 2020; Manniche et al., 2021; Rudan, 2023; Kabil et al., 2024). This study will compile various current reports and resources related to waste generated by the tourism sector and its subcategories, as well as waste management. The study will provide a brief work of the tourism sector's waste management policies within the context of a circular economy.

TOURISM INDUSTRY AND WASTE MANAGEMENT POLICIES

Within tourism waste management, waste types can be categorized as municipal and household waste. For example, waste generated by the tourism sector can be categorized as follows: (Sustainability Directory, 2025):

- Solid Waste in Tourism: This includes solid waste such as packaging, paper, plastic, and glass, belonging to tourists in areas they frequent, such as hotels and restaurants.
- Liquid Waste in Tourism: Sewage, runoff, and wastewater from hotels and resorts are considered tourism-related liquid waste.
- Gas Waste in Tourism: This refers to emission gases generated during travel in the tourism sector.
- Hazardous Waste in Tourism: This includes batteries, chemical cleaning products, and medical waste within tourism-related waste.

Data on waste management can be instructive in the tourism industry's transition to a circular economy. This study will examine statistics, indicators, and cases related to waste practices in the tourism industry.

It can be argued that waste management in the tourism sector is increasing due to the demands of sustainable tourism and environmentally conscious tourists. First, a look at the sustainable tourism market can be helpful in understanding the importance of waste management.

The goal of the Sustainable Tourism market is to offer a tourism experience that protects the environment by focusing on environmentally conscious tourist demand. The role and responsibility for solving environmental problems are outlined in the 2030 SDGs, and the tourism sector is also playing its part. Consumer demand is believed to be highly influential in the growth of sustainable tourism. For example, a 2023 survey indicated that 76 percent of tourists are interested in sustainable travel. The tourism sector is also implementing measures to reduce carbon emissions in line with the 2030 SDGs. Consequently, investments in sustainable tourism continue to increase. (Market Research. BIZ, 2024).

Table 1. Key factors of Sustainable Tourism Market

Key factor	**How they influence**
Market value	\$3.2 trillion (2023) \$11.2 trillion (2033) - expected
Market segment of tourism type	32% - Ecotourism is a leading segment
Market segment of service type	41% - Environmentally Friendly Destinations
Users	29% - Couple tourists travel
Leading region	42.1% - North America Leads
Opportunities	Future growth expectations are strong

Source: Market Research. BIZ, (2024)

According to Market Research BIZ (2024), some main key issues are shown in Table 1. It was seen that market values was recorded as 3.2 USD trillion and it was expected to reach 11.2 USD trillion by 2033. Eco-tourism seems a leading tourism market and North America seems as a leading region in sustainable tourism category.

Table 2. Market Size of Sustainable Tourism

Year	Market Value (USD Trillion)
2025	3.56
2026	4.06
2027	4.64
2028	5.31
2029	6.08
2030	6.97
2031	7.99
2032	9.17
2033	10.24
2034	11.39

Source: Precedence Research, 2025

In table 2, market value of sustainable tourism was shown between 2025 and 2034(expected). In 2025, the market value was recorded as 3.56 USD trillion and 11.39 USD trillion was expected by 2034 (Precedence Research, 2025).

Table 3. Regions in Sustainable Tourism

Regions	Position
Asia-Pacific	The Asia-Pacific region is projected to be the fastest-growing market in terms of sustainable tourism. Investments in sustainable tourism are increasing in the Asia-Pacific region.
China	China is increasing its investments in the sustainable tourism market. It prioritizes tourism focused on waste reduction and environmental protection.
Europe	There is an increase in the number of businesses in Europe adapting environmentally friendly and sustainable practices. Businesses in the tourism sector also support the green transformation towards sustainable practices (France, Germany and UK are leading countries for sustainable tourism).
Middle East and Africa	Regions such as the Middle East and Africa have also benefited from sustainable tourism-related sub-sectors.
UAE	The UAE is a leading country in the region, offering numerous options for sustainable tourism.

Source: Precedence Research, 2025

Table 3 shows the positions of some regions according to sustainable tourism pricing. Table 4 also shows the global sustainable tourism shares (%) of regions for the year 2024. As can be seen, while the European region is in a leading position in sustainable tourism, the Asia-Pacific region has a rapidly developing market.

Table 4. Share of Regions (2024) for Sustainable Tourism

Regions	Percent(%)
Asia-Pacific	20
Europe	31
Latin America	8
MEA	3
North America	38

Source: Precedence Research, 2025

Table 5. Travelers' Perception of Environmental Concern

Factors	Percent (%)
Excess waste	46
Wildlife damage	38
Overcrowding in destinations	30
Carbon emissions	29

Source: Liu, (2022a)

Table 5 shows the percentage distributions for some consumers' perceptions of environmental awareness and environmental protection during their travels. According to Liu (2022a), consumers are most concerned about excess waste (%46).

Table 6. Perception of Waste among Tourism Sector

Sustainability Issues	Percent (%)
Reducing energy consumption	83
Using environmentally friendly transport options	79
Reducing water usage	76
Reducing carbon footprint	69

Liu, (2022b)

Table 6, compiled from Liu's (2022b) data, presents the sustainable elements and the percentage shares of consumers. Accordingly, it can be said that consumers are most sensitive to energy reduction, environmentally friendly travel options, reducing water consumption, and reducing their carbon footprint.

Booking.com's 2021 sustainability report, based on a survey of approximately 30,000 travelers from 30 countries, identified reducing waste and energy consumption as key issues in sustainable tourism (Liu, 2022b).

Table 7. Conflicts in Sustainable Tourism

Country	Travel to nature (%)	Concern for nature (%)
Germany	52	18
China	46	38
France	45	22
United States	36	22
United Kingdom	33	23

Source: Brandt, 2022

Table 7 illustrates a fundamental dilemma in sustainable tourism. Consumers who demand sustainable tourism and those who are environmentally conscious should not be considered in the same category. According to Statista's global consumer survey data, while 52% of consumers in Germany want to travel in nature, only 18% are concerned about the environment (Brandt, 2022). The fact that travelers who want to travel in nature do not share the same environmental concerns can be considered a dilemma.

Statista Research Department (2025) gave recent statistics for perception of tourists' about environmental damage during the travel. As seen in table 8, UK tourists mostly concerned for flying's impact on environment rather than destination.

Table 8. Tourist Concern: Destination versus Flying

Age group	Destination	Flying
16-24	36	26
25-34	30	36
35-44	32	22
45-54	25	22
55 -upper	23	22

Source: Statista Research Department (2025)

It is important to consider that as the tourism and hospitality sector grows every year, the amount of waste also increases. Unfortunately, hoteliers release large amounts of waste into the environment every year, causing environmental pollution. Waste management is a serious issue for hotel managers. In this regard, avoiding the use of single-use plastic products can significantly reduce the amount of waste. With food waste management, waste can be collected and recycled in a way that causes less harm to the environment. Guest waste, however, is a challenging area for hotel managers. Guests should be clearly informed about what to do with their waste (Simran, 2025).

The increase in tourist travel after the COVID-19 period is causing a rise in concerns about tourism-related waste and environmental pollution. In particular, it would be beneficial for countries receiving the most international tourists to add new policies and strategies to their waste management policies to address waste generated by tourists or tourism activities. Table 9 shows the ranking of countries based through the most traveled place in the world as below:

Table 9. Most traveled countries POST-COVID-19

Countries	2019	2020	2021	2022	2023
Austria	31,88	15,09	12,73	26,22	30,91
China	65,73	7,97	5,66	-	-
France	90,91	41,68	48,4	93,2	100
Germany	39,56	12,45	11,69	28,46	34,8
Greece	31,35	7,37	14,71	27,84	32,74
Italy	64,51	25,19	26,89	49,81	57,25
Mexico	45,02	24,82	31,86	38,33	42,15
Spain	83,51	18,93	31,18	71,66	85,17
Turkey	51,19	15,89	29,93	50,45	55,16
UAE	21,56	7,17	11,48	22,65	28,15
UK	39,42	10,71	6,29	30,74	37,22
US	79,44	19,21	22,28	50,87	66,48

Source: WP Travel, (2025).

Table 9 shows the destination countries that hosted the most visitors in terms of post-COVID-19 visitor numbers. Accordingly, tourism-related waste and waste management issues can be considered more important in these countries. It can also be said that the tourism sector stands out in waste management for countries that host the most tourists in the world and have tourism-dependent income and economic systems.

CONCLUSION

The tourism industry, as one of the industries responsible for accelerating climate change and increasing environmental pollution, must also transition to a green transformation model. While studies on waste management in the tourism sector are noteworthy in the literature, it is also observed that tourists are increasingly inclined towards sustainable hotel and tourism practices. In waste management, tourism and hotel services generate different types and structures of waste depending on their sub-services. Different types of waste need to be separated and disposed of.

Accordingly, businesses can follow different methods in managing the waste they produce within the scope of tourism activities.

Some regions or countries in the world attract a higher number of tourists. An increase in the number of tourists will also increase the amount of tourism-related waste and environmental pollution. Therefore, waste management practices are even more important for countries with high tourist numbers. For countries that do not have active or successful waste management practices, urgent action plans should be developed and waste policies in the tourism sector should be reviewed (Perkumienė et al., 2023; Obersteiner et al., 2021; Phu et al., 2019; Wang et al., 2021; Shamshiry et al., 2011; Yusoff et al., 2022).

Regarding waste management policy recommendations for the tourism industry, the following can be stated:

- In densely populated countries, local governments should take more responsibility for waste management and cooperate with the tourism sector.
- In countries that regularly host a high number of tourists, legal responsibility for waste management should be increased for hotel and tourism professionals, and environmental legislation should be revised.
- Training should be provided to tourism sector employees to raise their environmental awareness.
- Tourists should be informed about the waste they leave behind in the countries they visit.
- Sustainable transformation in the tourism sector should be accelerated.
- The necessary infrastructure and financial support for waste management should be accelerated.

REFERENCES

Abdulkadir, A., Ally, B., Remmen, A., Hirsbak, S., & Salukele, F. (2025). Circular Economy Framework for Plastic Waste Management: A Case Study from Coastal Hotels in Zanzibar. *Resources*, *14*(11), 175. DOI: 10.3390/resources14110175

Ahmad, N., & Ma, X. (2021). How does tourism development affect environmental pollution? Tourism Economics, 28(6), 1453-1479. (Original work published 2022)DOI: 10.1177/13548166211000480

Axhami, M., Ndou, V., Milo, V., & Scorrano, P. (2023). Creating Value via the Circular Economy: Practices in the Tourism Sector. *Administrative Sciences*, *13*(7), 166. DOI: 10.3390/admsci13070166

Azam, M., Mahmudul Alam, M., & Haroon Hafeez, M. (2018). Md Mahmudul Alam, Muhammad Haroon Hafeez, (2018). Effect of tourism on environmental pollution: Further evidence from Malaysia, Singapore and Thailand. *Journal of Cleaner Production*, *190*, 330–338. DOI: 10.1016/j.jclepro.2018.04.168

Brandt, M. (2022). Nature Lovers Don't Equal Sustainable Travelers, https://www.statista.com/chart/27611/survey-on-the-importance-of-nature-and-sustainability-when-travelling/?srsltid=AfmBOopxJjqKJcO4A_Fu05VrlnraT9J2L7pxM66wn0pL598SAd-8bSKk

Camilleri, M. A. (2021). Sustainable Production and Consumption of Food. Mise-en-Place Circular Economy Policies and Waste Management Practices in Tourism Cities. *Sustainability (Basel)*, *13*(17), 9986. DOI: 10.3390/su13179986

Chiwaridzo, O. T., & Han, J. (2025). Circular economy strategies for waste management: The mediating role of green tourism supply chain management in environmental management and energy sustainability within the Zimbabwean tourism sector. Waste Management & Research. *Waste Management & Research*, *0*(0), 0734242X251385933. Advance online publication. DOI: 10.1177/0734242X251385933 PMID: 41217274

Falcone, P. M. (2019). Tourism-Based Circular Economy in Salento (South Italy): A SWOT-ANP Analysis. *Social Sciences (Basel, Switzerland)*, *8*(7), 216. DOI: 10.3390/socsci8070216

Gao, J., & Zhang, L. (2021). Exploring the dynamic linkages between tourism growth and environmental pollution: New evidence from the Mediterranean countries. *Current Issues in Tourism*, *24*(1), 49–65. DOI: 10.1080/13683500.2019.1688767

Giurea, R., Precazzini, I., Ionescu, G., Ragazzi, M., & Schiavon, M. (2022, August 17). Ilaria Precazzini, Gabriela Ionescu, Marco Ragazzi, Marco Schiavon, (2022). Circular economy, waste and energy management for a sustainable agro-tourism. *AIP Conference Proceedings*, *2437*(1), 020097. DOI: 10.1063/5.0093290

Islam, N. F., Gogoi, B., Saikia, R., Yousaf, B., Narayan, M., & Sarma, H. (2024). Bhoirob Gogoi, Rimon Saikia, Balal Yousaf, Mahesh Narayan, Hemen Sarma, (2024). Encouraging circular economy and sustainable environmental practices by addressing waste management and biomass energy production. *Regional Sustainability*, *5*(4), 100174. DOI: 10.1016/j.regsus.2024.100174

Kabil, M., Rahmat, A. F., Hegedüs, M., & Dávid, L. D. (2024). Circular Economy and Tourism: A Bibliometric Journey Through Scholarly Discourse. *Journal of Circular Economy*, *2*(1). Advance online publication. DOI: 10.55845/HGWO7144

Katircioglu, S. T. (2014). International tourism, energy consumption, and environmental pollution: The case of Turkey. *Renewable & Sustainable Energy Reviews*, *36*, 180–187. DOI: 10.1016/j.rser.2014.04.058

Laureti, L., Costantiello, A., Anobile, F., Leogrande, A., & Magazzino, C. (2024). Waste Management and Innovation: Insights from Europe. *Recycling*, *9*(5), 82. DOI: 10.3390/recycling9050082

Liu, S. (2022a). Excess waste a key sustainability concern for travelers, https://www.statista.com/chart/28149/excess-waste-a-key-sustainability-concern/?srsltid=AfmBOopWKGMdBlm-QyBp1jdFIxj165U2rb2TDaCULJ5adtTpggFuNPTl

Liu, S. (2022b). Consumers want to travel more sustainably, https://www.statista.com/chart/27980/travel-more-sustainably/?srsltid=AfmBOoqU2vAfdpFurB6nUF-_f4NHdn0VznW6ep5edaSX_tuOuG_cKpat

Manniche, J., Larsen, K. T., & Broegaard, R. B. (2021). The circular economy in tourism: Transition perspectives for business and research. *Scandinavian Journal of Hospitality and Tourism*, *21*(3), 247–264. DOI: 10.1080/15022250.2021.1921020

Market Research. BIZ, (2024). Sustainable Tourism Market Report By Tourism Type (Ecotourism, Adventure Tourism, Cultural Tourism, Wildlife Tourism, Community-based Tourism, Green Tourism, Agro-tourism), By Service Type (Accommodation [Eco-friendly Hotels, Green Resorts, Sustainable Lodges], Transportation [Eco-friendly Transport, Public Transport, Bicycle Rentals], Activities and Tours, Food and Beverage, By End User, By Region and Companies - Industry Segment Outlook, Market Assessment, Competition Scenario, Trends and Forecast 2024-2033, https://marketresearch.biz/report/sustainable-tourism-market/

Metilelu, O. O., & Michael, O. (2022). Adeniyi, Matthew I. Ekum, (2022). Modelling the Dynamic Effect of Environmental Pollution on Coastal Tourism. *Scientific African*, *17*, e01364. DOI: 10.1016/j.sciaf.2022.e01364

Munonye, W. C. (2025). Integrating Circular Economy Principles in Business Strategies: A Policy-Driven Approach. *Circular Economy and Sustainability*, *5*(3), 1887–1896. DOI: 10.1007/s43615-025-00523-5

Obersteiner, G., Gollnow, S., & Eriksson, M. (2021). Carbon footprint reduction potential of waste management strategies in tourism. *Environmental Development*, *39*, 100617. DOI: 10.1016/j.envdev.2021.100617 PMID: 34513580

OECD. (2019). *Waste Management and the Circular Economy in Selected OECD Countries: Evidence from Environmental Performance Reviews, OECD Environmental Performance Reviews*. OECD Publishing., DOI: 10.1787/9789264309395-

Perkumienė, D., Atalay, A., Safaa, L., & Grigienė, J. (2023). Sustainable Waste Management for Clean and Safe Environments in the Recreation and Tourism Sector: A Case Study of Lithuania, Turkey and Morocco. *Recycling*, *8*(4), 56. DOI: 10.3390/recycling8040056

Precedence Research. (2025). Sustainable Tourism Market Size, Share, and Trends 2025 to 2034, Sustainable Tourism Market (By Type: Island Tourism, Mountain Tourism, Coastal Tourism; By Application: Group, Family, Solo, Couples; By Age group: Millennial, Baby Boomers, Generation X, Silver hair) - Global Industry Analysis, Size, Trends, Leading Companies, Regional Outlook, and Forecast 2025 to 2034, https://www.precedenceresearch.com/sustainable-tourism-market

Rodríguez, C., Florido, C., & Jacob, M. (2020). Circular Economy Contributions to the Tourism Sector: A Critical Literature Review. *Sustainability (Basel)*, *12*(11), 4338. DOI: 10.3390/su12114338

Rodríguez, C., Florido, C., & Jacob, M. (2020). Circular Economy Contributions to the Tourism Sector: A Critical Literature Review. *Sustainability (Basel)*, *12*(11), 4338. DOI: 10.3390/su12114338

Rudan, E. (2023). Circular Economy of Cultural Heritage—Possibility to Create a New Tourism Product through Adaptive Reuse. *Journal of Risk and Financial Management*, *16*(3), 196. DOI: 10.3390/jrfm16030196

Shamshiry, Elmira, Nadi, Behzad, Bin Mokhtar, Mazlin, Komoo, Ibrahim, Saadiah Hashim, Halimaton, Yahaya, Nadzri, Integrated Models for Solid Waste Management in Tourism Regions: Langkawi Island, Malaysia, Journal of Environmental and Public Health, 2011, 709549, 5 pages, 2011. DOI: 10.1155/2011/709549

Simran, S. (2025). How To Reduce Cost And Waste In A Hotel - Guide For Hotel Owner, https://www.bdtask.com/blog/how-to-reduce-cost-and-waste-in-a-hotel

Sørensen, F., & Bærenholdt, J. O. (2020). Jørgen Ole Bærenholdt, (2020). Tourist practices in the circular economy. *Annals of Tourism Research*, *85*, 103027. Advance online publication. DOI: 10.1016/j.annals.2020.103027

Statista Research Department. (2025). Concerns over the environmental impact of tourism among adults in the United Kingdom as of January 2025, by age, https://www.statista.com/statistics/1135573/tourism-sustainability-concerns-uk-by-age/?srsltid=AfmBOopsQ7nzfQK_Man6ttR9TJzL5MYhEHHHjWIkV0DQ0XChQ8gbnnck

Strippoli, R., Gallucci, T., & Ingrao, C. (2024). Circular economy and sustainable development in the tourism sector – An overview of the truly-effective strategies and related benefits. *Heliyon*, *10*(17), e36801. https://www.cell.com/heliyon/fulltext/S2405-8440(24)12832-0. DOI: 10.1016/j.heliyon.2024.e36801 PMID: 39281586

Sustainability Directory. (2025). Tourism Waste Management, https://pollution.sustainability-directory.com/term/tourism-waste-management/#:~:text=Types%20of%20Tourism%20Waste&text=Solid%20Waste%20%E2%86%92%20This%20is,%2C%20attractions%2C%20and%20public%20spaces

Travel, W. P. (2025). Global Tourism Statistics 2024:Trends and Forecasts, https://wptravel.io/global-tourism-industry-statistics/#h-global-tourism-statistics-2019

Wang, K. C. M., Lee, K. E., & Mokhtar, M. (2021). Solid Waste Management in Small Tourism Islands: An Evolutionary Governance Approach. *Sustainability (Basel)*, *13*(11), 5896. DOI: 10.3390/su13115896

Xu, A., Wang, C., Tang, D., & Ye, W. (2022). Tourism circular economy: Identification and measurement of tourism industry ecologization. *Ecological Indicators*, *144*, 109476. DOI: 10.1016/j.ecolind.2022.109476

Yıldırım, D. Ç., Ektiren, B., & Erdoğan, F. (2023). Burak Ektiren, Fatma Erdoğan, (2023). The threshold effect of tourism, growth, and population interactions on environmental pollution in Mediterranean countries. *Tourism Management Perspectives*, *47*, 101118. DOI: 10.1016/j.tmp.2023.101118

Yıldırım, S. (2025). Unveiling Circularity: Challenges and Barriers in Transitioning to Circular Practices. In Singh, P., Daga, S., Yadav, K., & Mishra, V. (Eds.), *Circular Economy and Environmental Resilience*. Springer., DOI: 10.1007/978-3-031-93091-1_8

Yıldırım, S., Yıldırım, D. Ç., Aydın, K., & Erdoğan, F. (2021). Regime-dependent effect of tourism on carbon emissions in the Mediterranean countries. *Environmental Science and Pollution Research International, 28*(39), 54766–54780. DOI: 10.1007/s11356-021-14391-7 PMID: 34014485

Yusoff, M. S.. (2022). Solid Waste Management in the Tourism Industry. In Wang, L. K., Wang, M. H. S., & Hung, Y. T. (Eds.), *Solid Waste Engineering and Management. Handbook of Environmental Engineering* (Vol. 25). Springer., DOI: 10.1007/978-3-030-96989-9_1

Zhang, L., & Gao, J. (2016). Exploring the effects of international tourism on China's economic growth, energy consumption and environmental pollution: Evidence from a regional panel analysis. *Renewable & Sustainable Energy Reviews, 53*, 225–234. DOI: 10.1016/j.rser.2015.08.040

ADDITIONAL READING

Bostancı, S. H., & Yıldırım, S. (2021). Sustainable Communities vs. Climate Refugees: Two Opposite Results of Climate Change. In Popescu, C. (Ed.), *Handbook of Research on Novel Practices and Current Successes in Achieving the Sustainable Development Goals* (pp. 298–319). IGI Global., DOI: 10.4018/978-1-7998-8426-2.ch015

Bostancı, S. H., & Yıldırım, S. (2022). The Role of Municipalities in Achieving Water Security: The Case of Turkey. In Castanho, R. (Ed.), *Handbook of Research on Sustainable Development Goals, Climate Change, and Digitalization* (pp. 268–286). IGI Global., DOI: 10.4018/978-1-7998-8482-8.ch017

Sevik, N., & Yıldırım, S. (2023). Sustainable Enterprises Post COVID-19 Pandemic: The Rise of Sustainable Fashion Enterprises. In Castanho, R. (Ed.), *Handbook of Research on Current Advances and Challenges of Borderlands, Migration, and Geopolitics* (pp. 343–359). IGI Global., DOI: 10.4018/978-1-6684-7020-6.ch017

Yildirim, S. (2020). The consumer role for sustainable development: how consumers contribute sustainable development goals. In Chkoniya, V., Madsen, A. O., & Bukhrashvili, P. (Eds.), *Anthropological Approaches to Understanding Consumption Patterns and Consumer Behavior* (pp. 325–341). IGI Global. DOI: 10.4018/978-1-7998-3115-0.ch018

Yıldırım, S. (2021). Do green women influencers spur sustainable consumption patterns? Descriptive evidences from social media influencers. *Ecofeminism and Climate Change*, *2*(4), 198–210. DOI: 10.1108/EFCC-02-2021-0003

Yildirim, S. (2021). 3R Principle in Household Energy Consumption Patterns: A Case of Turkish Households. In Danish, M., & Senjyu, T. (Eds.), *Eco-Friendly Energy Processes and Technologies for Achieving Sustainable Development* (pp. 88–107). IGI Global. DOI: 10.4018/978-1-7998-4915-5.ch005

Yıldırım, S., & Bostancı, S. H. (2021). The efficiency of e-government portal management from a citizen perspective: Evidences from Turkey, *World Journal of Science. Technology and Sustainable Development*, *18*(3), 259–273. DOI: 10.1108/WJSTSD-04-2021-0049

Yıldırım, S., & Candan, B. (2015). Segmentation of green product buyers based on their personal values and consumption values. *Environmental Values*, *24*(5), 641–661. DOI: 10.3197/096327115X14384223590212

Yildirim, S., Demirtas, I., & Yildirim, D. C. (2022). A Review of Alternative Economic Approaches to Achieve Sustainable Development: The Rising Digitalization and Degrowth Post COVID-19. In Castanho, R. (Ed.), *Handbook of Research on Sustainable Development Goals, Climate Change, and Digitalization* (pp. 288–307). IGI Global. DOI: 10.4018/978-1-7998-8482-8.ch018

Yıldırım, S., & Kantarcı, T. (2022). A Review on Sustainability Policies of Businesses: Recycling and Waste Reduction. *Journal of Recycling Economy & Sustainability Policy*, *1*(1), 1–9. https://respjournal.com/index.php/pub/article/view/1

Yildirim, S., Kaplan, M., & Yildirim, D. C. (2023). Climate Change and the Ocean Economy: A Review on OECD's Ocean Economy. In Pego, A. (Ed.), *Handbook of Research on Bioeconomy and Economic Ecosystems* (pp. 212–232). IGI Global., DOI: 10.4018/978-1-6684-8879-9.ch012

Yıldırım, S., Sevik, N., Kandpal, V., & Yıldırım, D. C. (2024). The Role of Green Brands on Achieving 2030 Sustainable Development Goals (2030 SDGs). In Kankaew, K., Chaudhary, S., & Widtayakornbundit, S. (Eds.), *Contemporary Management and Global Leadership for Sustainability* (pp. 141–162). IGI Global., DOI: 10.4018/979-8-3693-1273-5.ch009

Yıldırım, S., & Vasilev, V. (2024). The Future of Green Collar Workers: Green Transition in Employment. In Yıldırım, S., Yıldırım, D., Demirtaş, I., & Kandpal, V. (Eds.), *Green Transition Impacts on the Economy, Society, and Environment* (pp. 1–16). IGI Global., DOI: 10.4018/979-8-3693-3985-5.ch001

Yıldırım, S., & Yıldırım, D. Ç. (2020). Achieving Sustainable Development Through a Green Economy Approach. In Patti, S., & Trizzino, G. (Eds.), *Advanced Integrated Approaches to Environmental Economics and Policy: Emerging Research and Opportunities* (pp. 1–22). IGI Global. DOI: 10.4018/978-1-5225-9562-5.ch001

Yıldırım, S., & Yıldırım, D. C. (2022). Achieving Seafood Security in the Mediterranean Region: A Case of Turkey. In I. Management Association (Ed.), Research Anthology on Ecosystem Conservation and Preserving Biodiversity (pp. 1780-1800). IGI Global. DOI: 10.4018/978-1-6684-5678-1.ch086

Yıldırım, S., Yıldırım, D. C., Demirtaş, I., & Kandpal, V. (Eds.). (2024). *Green Transition Impacts on the Economy, Society, and Environment*. IGI Global., DOI: 10.4018/979-8-3693-3985-5

Yildirim, S., Yildirim, D. C., & Erdogan, F. (2022). The Sustainability Level of Agricultural Food Markets: Selected Indicators From Mediterranean Countries. In Castanho, R. (Ed.), *Analyzing Sustainability in Peripheral, Ultra-Peripheral, and Low-Density Regions* (pp. 131–155). IGI Global., DOI: 10.4018/978-1-6684-4548-8.ch008

Yildirim, S., Yildirim, D. C., & Erdogan, S. (2024). Energy Security Dilemma in OECD Countries: The COVID-19 Pandemic, Russia-Ukraine War, and Energy Issues. In R. Castanho (Ed.), Green Economy and Renewable Energy Transitions for Sustainable Development (pp. 83-105). IGI Global. DOI: 10.4018/979-8-3693-1297-1.ch005

Yildirim, S., Yildirim, D. C., & Esen, Ö. (2023). Environmental Protection and Environmental Protection Expenditure in Developing Countries: A Case of Turkey. In Castanho, R. (Ed.), *Handbook of Research on Current Advances and Challenges of Borderlands, Migration, and Geopolitics* (pp. 100–119). IGI Global., DOI: 10.4018/978-1-6684-7020-6.ch005

Chapter 4
A Review of Key Indicators for Waste Management:
Turkey Case

Seda Yıldırım
https://orcid.org/0000-0003-4367-6652
Tekirdag Namık Kemal University, Turkey

Durmuş Çağrı Yıldırım
Tekirdag Namık Kemal University, Turkey

ABSTRACT

Waste management has become a prominent application with the green economy and circular economy. The most important circular application area of the 21st century can be considered as elements related to waste management processes. Innovative approaches and alternative policies are being developed for the improvement of waste management on a global scale. There may be differences between the approaches and practices adopted by each country for waste management process. However, generally, the assistance of smart technologies and the inclusion of the public in the waste management process can be considered as common policies. Technological infrastructure and integration will yield more efficient results in the long term. At this point, the aim of the study is to examine

DOI: 10.4018/979-8-3373-9685-9.ch004

the cyclical applications in waste management. Turkey was selected as a sample country. Turkey is a developing country that has taken important steps in combating climate change. In this regard, a sample country review will be conducted by presenting useful information on waste management and circular applications.

INTRODUCTION

Combating environmental pollution and protecting resources to combat climate change have become priorities for policymakers. Due to their exacerbating impact on environmental pollution, waste management and recycling are among the most fundamental sustainable practices. Waste management and recycling are transforming the economy through a new economic model: the circular economic model (Bellezoni et al., 2022; Yang et al., 2023; Christis et al., 2019; Serrano et al., 2021; Romero-Perdomo et al., 2022; Möslinger et al., 2023; Yildirim et al., 2025; Yildirim and Bostanci, 2025).

The circular economy can be considered a practical approach to the green economy and sustainable economy. Initially, the concepts of green economy and sustainable economy remained broad and abstract. While businesses and many industries struggle to adopt realistic and successful sustainable practices during the green transformation, the circular economy phenomenon has brought a new perspective. In this context, circular economic practices offer practices that are more easily understood and adopted by industry and businesses, aligning with sustainable development. Waste management and recycling can be considered two fundamental circular practices that form the heart of the circular economy, creating hope for accessibility for a sustainable future (Klein et al., 2020; Awan and Sroufe, 2022; Yıldırım and Kantarcı, 2022; Suárez-Eiroa et al., 2019; Puntillo, 2023; Mandpe et al., 2022; Schroeder et al., 2019).

Figure 1. Circular Economy Model

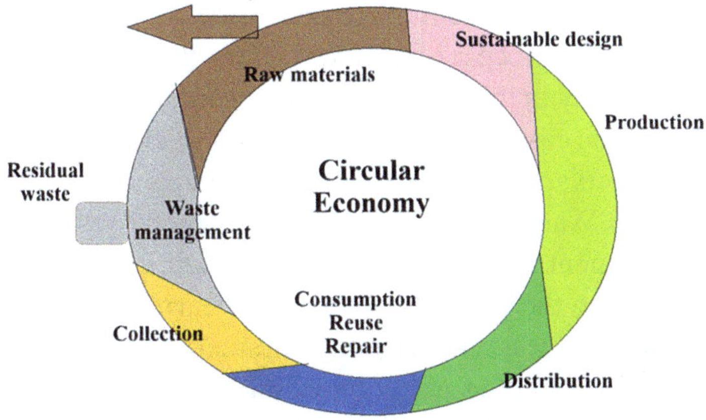

Source: European Parliament, 2023

Figure 1 illustrates the role of waste management and recycling practices within a circular economy. In the circular economy model, used products and raw materials are reintegrated into production through waste management and recycling.

The circular economy offers an economic model that prevents resource waste. It provides a return to the waste and raw material-resource relationship to halt and prevent environmental damage resulting from the linear economic model (OECD, 2020). It can be argued that environmental damage has increased as countries seeking post-war recovery through a linear economy have accelerated their prosperity. Increased resource use, consumption, and production, along with post-consumption waste, have exacerbated environmental pollution in the long term. Furthermore, threats such as the limited availability of natural resources and the long-term absence of necessary resources for future generations have emerged. The circular economy aims to use and conserve resources efficiently. Waste reduction after resource consumption and the transformation of waste into resources are also among the contributions of the circular economy (European Investment Bank, 2023).

The circular economy model has expanded waste management and recycling practices, bringing the concept of a digital plastic economy to the literature. The fact that plastic pollution poses a more difficult environmental challenge has made recycling plastic waste a mandatory practice. Recycling plastic waste offers an opportunity to reduce the amount of plastic waste accumulating in the environment (Nixon et al., 2024). Nixon et al. (2024) focused on the circular process of plastic waste within the framework of a plastic waste economy in their study. In this regard, they examined some industrial application cases. It can be said that smart city management has achieved better circular practices thanks to advances in the internet and smart technologies. Waste collection and disposal processes are critical for smart city management and sustainable cities (Chen, 2022). Chen (2022) conducted a study on the usability of machine learning models for both waste management and waste management in smart cities. It was observed that waste collection points, such as recycling bins and dumpsters, have become more effective with AI and machine learning. In recent years, European countries have also begun to grapple with increasing waste volumes. With Europe's increasing dependence on both raw materials, materials, and energy, waste management and recycling have become critical practices. Europe plans to reduce its material-dependent economy through recycling (Ragossnig and Schneider, 2019). Van Caneghem et al. (2019) conducted a study on waste-to-energy conversion. They examined the potential use of municipal solid waste for energy recycling. Essentially, the conversion of waste to energy is a recycling practice. Non-recyclable waste is also disposed of through incineration.

Waste management effectiveness should be considered as a practice that can also impact human health in the long term. Improper waste management pollutes the environment and causes harmful toxic substances to accumulate in the environment, potentially harming both animals and humans. Sustainable production and consumption—Goal 12—should not be considered merely a draft mandate, but rather a practical goal to be achieved. Therefore, waste management and recycling should be carried out correctly, considering green technology and the environment, to create a sustainable future for both environmental and human health (Sasmoko et al., 2022). Within the circular economy, resource management and waste management are considered as processes that work together in harmony. Effective management of resource input and output provides benefits such as sustainable production, zero waste, and the efficient use of resources. Circular economy and recycling practices can be pursued in different sectors. For example, recycling can be pursued in the construction sector for demolition and construction materials (Liu et al., 2021). Mikhno et al. (2023) discussed the importance of recycling practices for construction materials and waste. They provided an example of a study on the circular economy and the construction sector in Ukraine. Ding et al. (2023) gave evidences for the importance of recycling practices in construction sector in the context of sustainable cities (Ding et al., 2023). As can be seen in recent literature, researchers also expect the circular economy to transform urban development. Recycling of building materials is seen as important for both waste management and the rapid pace of urbanization in circular practices. On the other side, there are many circular practices and waste management policies among the various industries.

Waste pollution belonged to each type of industry and sector in the world. Accordingly, waste management policy should be launched for each type of industry to achieve sustainable life. In densely populated and crowded cities and regions, municipalities have a greater role and responsibility in waste management. In crowded cities and regions, municipal recycling and waste management practices can be made more effective by developing new strategies (Rolewicz-Kalińska et al., 2020)

Aim and Design

This study aims to examine waste management practices and data using the example of Turkey. In other words, waste management indicators will be examined, along with some of the current circular waste management practices. This study will analyze waste management for Turkey using open access data from the TURKSTAT (2025)'s database.

TURKEY CASE: INDICATORS AND REPORTS ON WASTE MANAGEMENT AND PRACTICES

Turkey is pursuing significant sustainable public and environmental policies within the scope of the 2030 SDGs. The Zero Waste project is one of the important projects pioneered in Turkey. The Zero Waste project is a key public policy encompassing waste management, waste prevention, and recycling practices crucial for sustainable development. Waste management is considered an important and growing area of practice in Turkey. The First Lady Emine Erdogan launched Zero Waste Project in 2017 and Zero Waste Foundation was established in Turkey in 2023 (Zero Waste Foundation, n.d). Zero Waste Project was awarded by UNPD Türkiye in 2021. Accordingly, this project is so important in fight against to climate change and environmental pollution (UNPD, Türkiye, 2021).

The latest TURKSTAT (2025) indicators regarding waste management in Turkey were utilized. Summary tables below present the latest developments in waste management in Turkey.

Table 1. Key Indicators for Waste Management in Turkey

	2022			2024		
Indicators	**Total**	**Hazardous**	**Non-H**	**Total**	**Hazardous**	**Non-H**
Waste Generated(amount)	27 969 021	5 439 883	22 529 139	24 350 094	4 191 675	20 158 419
Treatment	Total	Hazardous	Non-H	Total	Hazardous	Non-H
Recovered-Facility	4 056 403	2 217 023	1 839 379			
Sold/Sent-Waste Treatment Facilities	17 678 298	1 362 552	16 315 746	17 078 892	1 331 656	15 747 236
Used As Filling Material And For Reclamation	93 222	0	93 222			
Co-İncinerated/ İncinerated	569 737	50 498	519 239	628 305	45 737	582 568
Collected By Municipality/Organized İndustrial Zone	857 494	1 600	855 894	594 566	334	594 232
Sent To Controlled Landfill Sites	3 767 240	1 792 080	1 975 160	2 623 473	1 501 037	1 122 436
Stored Within The Establishment	924 356	16 130	908 226	449 887	9 817	440 071
Disposed Of By Other Methods	22 271	0	22 271	17 277	0	17 277

Source: TURKSTAT, 2025a

Table 1 shows some key indicators for Waste Management in Turkey based on TURKSTAT 2025's recent report on waste management. These variables belonged to waste management in 2024 in Turkey. Table 1 shows hazardous and non-hazardous waste amounts and numbers separately.

Table 2. Amount Of Waste Generated among various sectors in Turkey

Sectors	2022	2024
Manufacturing	27 969 021	24 350 094
Food, Beverage And Tobacco Products	1 385 195	1 632 066
Textiles, Wearing Apparel And Leather Products	824 364	737 691
Wood products(Except Furniture)	426 438	262 254
Pulp, Paper products	1 346 807	1 306 515
Coke, Refined Petroleum Products	67 194	76 244
Chemicals, and chemical products	6 129 198	5 052 112
Other Non-Metallic Mineral Products	1 275 497	1 300 236
Basic Metals etc products	14 376 602	11 705 716
Computur, Electronic Products Etc	1 904 585	2 060 803
Furniture, and etc products	233 141	216 457

Source: TURKSTAT, 2025b

Table 2 shows the Amount Of Waste Generated among various sectors in Turkey. There are numbers for 2022 and 2024. It can be seen that there are many sectors and industries for waste management. Some sectors have higher numbers of Waste Generated in 2024 rather than 2022.

Table 3. Amount (Tonnes) of Waste

Indicators	2022	2024
Total	348 038	353 222
Sent to waste treatment facilities	235 585	243 597
Disposal in municipal dumping sites	51 034	58 129
Dumping onto land	9 810	5 308
Land treatment resulting in benefit to agriculture or ecological improvement	19 766	159
Temporarily stored	30 290	45 673
Other	1 553	357

Source: TURKSTAT, 2025c

Table 3 gives the Amount (Tonnes) of Waste in Turkey comparing 2022 and 2024. Table 4 and Table 5 are all give detailed waste indicators of municipalities in Turkey. There are city details for municipalities among the waste management and indicators in 2024.

Table 4. Waste Indicators of Municipality in Turkey (2024)

City Info	Municipalities (number)	Waste(amount)	Municipality's dumping sites Number	Municipality's dumping sites waste (Tonnes)	Waste treatment facilities Number	Waste treatment facilities waste (Tonnes)
Türkiye	1 392	32 279 224	507	3 513 303	1 020	28 685 210
Adana	16	760 976	0	0	16	759 776
Adıyaman	22	190 851	22	190 086	1	765
Afyonkarahisar	60	198 714	20	18 186	60	178 694
Ağrı	12	160 490	11	139 327	2	21 163
Amasya	8	119 450	1	1 250	6	117 923
Ankara	26	2 101 748	4	10 244	24	2 091 504
Antalya	20	1 514 739	3	8 285	18	1 506 454
Artvin	9	65 030	9	63 625	4	1 405
Aydın	18	506 708	0	0	17	506 703
Balıkesir	21	476 610	1	13 866	19	462 744
Bilecik	11	70 440	0	0	11	70 440
Bingöl	11	57 333	7	7 450	4	49 883
Bitlis	13	84 934	5	12 185	8	72 749
Bolu	12	109 713	10	30 140	3	79 573
Burdur	15	132 315	9	33 263	9	99 052
Bursa	18	1 278 248	3	11 140	17	1 267 108
Çanakkale	23	275 407	14	161 174	15	114 233
Çankırı	17	61 385	2	2 750	16	58 635
Çorum	16	139 778	4	7 110	11	131 228
Denizli	20	351 766	5	48 822	14	302 294
Diyarbakır	18	731 673	17	731 274	3	399
Edirne	16	146 073	4	5 804	12	131 517
Elazığ	20	174 098	16	39 392	5	134 706
Erzincan	15	82 025	8	15 924	8	64 641
Erzurum	21	229 331	17	80 867	4	122 083
Eskişehir	15	344 763	7	18 721	15	326 042
Gaziantep	10	751 848	0	0	10	751 848
Giresun	24	129 496	7	20 102	17	109 394
Gümüşhane	14	24 227	1	1 310	13	22 917

continued on following page

Table 4. Continued

City Info	Municipalities (number)	Waste(amount)	Municipality's dumping sites		Waste treatment facilities	
			Number	waste (Tonnes)	Number	waste (Tonnes)
Hakkari	8	64 570	0	0	8	64 570
Hatay	16	810 189	4	84 901	14	725 288
Isparta	22	168 563	11	38 991	18	129 572
Mersin	14	686 135	1	486	13	685 650
İstanbul	40	7 024 241	0	0	40	7 024 241
İzmir	31	2 009 256	0	0	31	2 006 936
Kars	9	74 514	9	73 420	1	1 094
Kastamonu	20	84 514	1	1 278	20	83 237
Kayseri	17	471 974	8	72 512	11	399 462
Kırklareli	21	119 468	4	5 732	19	113 736
Kırşehir	10	82 979	8	27 335	5	55 644
Kocaeli	13	754 312	0	0	13	734 312
Konya	32	919 745	23	257 131	22	662 614
Kütahya	28	154 751	10	9 745	28	144 324
Malatya	14	224 501	6	28 771	7	195 730
Manisa	18	586 890	0	0	17	577 350
Kahramanmaraş	11	356 435	4	101 200	8	255 235
Mardin	11	243 842	2	18 652	8	225 190
Muğla	14	735 310	2	205 230	13	530 080
Muş	22	84 023	21	39 273	2	44 750
Nevşehir	23	107 823	6	7 527	23	100 296
Niğde	29	113 969	1	800	28	113 169
Ordu	20	233 783	0	0	20	233 783
Rize	18	76 772	1	4 750	18	72 022
Sakarya	17	397 976	0	0	17	397 856
Samsun	18	414 427	0	0	18	414 427
Siirt	12	85 289	11	40 398	1	44 892
Sinop	9	67 944	3	23 650	6	44 294
Sivas	24	186 391	21	70 708	8	115 683
Tekirdağ	12	515 336	0	0	11	515 336
Tokat	40	177 321	16	25 161	29	150 359
Trabzon	19	225 668	0	0	19	225 623
Tunceli	9	27 030	8	24 742	5	2 288
Şanlıurfa	14	551 767	4	73 062	11	478 705
Uşak	11	119 438	9	19 024	6	100 414
Van	14	419 673	2	10 600	11	409 073

continued on following page

Table 4. Continued

City Info	Municipalities (number)	Waste(amount)	Municipality's dumping sites		Waste treatment facilities	
			Number	waste (Tonnes)	Number	waste (Tonnes)
Yozgat	36	127 910	31	57 449	9	70 461
Zonguldak	25	162 082	0	0	25	161 668
Aksaray	22	102 917	8	9 473	14	93 444
Bayburt	5	19 942	2	1 490	3	18 452
Karaman	11	65 228	6	7 078	6	55 990
Kırıkkale	11	78 349	7	10 548	5	67 801
Batman	11	173 737	11	173 590	1	147
Şırnak	19	223 324	19	220 674	3	1 950
Bartın	8	73 336	8	66 397	6	6 939
Ardahan	7	27 066	6	9 820	2	17 246
Iğdır	7	50 244	2	8 900	7	41 344
Yalova	14	123 717	1	5 360	14	117 427
Karabük	7	63 086	0	0	7	63 086
Kilis	4	72 659	1	3 000	3	69 659
Osmaniye	14	148 514	1	2 000	14	146 514
Düzce	10	120 127	1	150	10	119 977

Source: TURKSTAT, 2025d

As seen in Table 4 and Table 5, higher population-city should include higher numbers of municipality waste service to fight against to environmental pollution.

Table 5. Other Indicators for waste in Municipality 2024

	municipal population	municipalities(number)	Municipality waste services	Population	Amount of waste collected(amount)
Türkiye	81 253 846	1 401	1 392	80 793 311	32 279 224
Adana	2 280 484	16	16	2 267 381	760 976
Adıyaman	452 498	23	22	447 786	190 851
Afyonkarahisar	592 836	60	60	590 629	198 714
Ağrı	316 819	12	12	316 401	160 490
Amasya	260 964	8	8	259 716	119 450
Ankara	5 864 049	26	26	5 863 519	2 101 748
Antalya	2 722 103	20	20	2 722 103	1 514 739

continued on following page

Table 5. Continued

	municipal population	municipalities(number)	Municipality waste services	Population	Amount of waste collected(amount)
Artvin	108 825	9	9	107 396	65 030
Aydın	1 165 943	18	18	1 156 776	506 708
Balıkesir	1 276 096	21	21	1 275 269	476 610
Bilecik	197 959	11	11	197 839	70 440
Bingöl	203 991	11	11	201 896	57 333
Bitlis	259 620	13	13	259 310	84 934
Bolu	247 209	12	12	247 209	109 713
Burdur	204 497	15	15	203 709	132 315
Bursa	3 238 618	18	18	3 237 665	1 278 248
Çanakkale	424 768	24	23	415 311	275 407
Çankırı	150 527	17	17	148 655	61 385
Çorum	398 847	16	16	397 928	139 778
Denizli	1 061 371	20	20	1 056 609	351 766
Diyarbakır	1 833 684	18	18	1 792 494	731 673
Edirne	338 887	16	16	335 595	146 073
Elazığ	516 373	20	20	513 688	174 098
Erzincan	194 732	15	15	194 524	82 025
Erzurum	745 005	21	21	704 494	229 331
Eskişehir	921 630	15	15	918 915	344 763
Gaziantep	2 193 363	10	10	2 193 363	751 848
Giresun	326 268	24	24	316 986	129 496
Gümüşhane	109 541	15	14	106 805	24 227
Hakkari	195 824	8	8	195 385	64 570
Hatay	1 562 185	16	16	1 560 670	810 189
Isparta	358 586	22	22	358 324	168 563
Mersin	1 954 279	14	14	1 954 279	686 135
İstanbul	15 701 602	40	40	15 701 602	7 024 241
İzmir	4 493 242	31	31	4 488 902	2 009 256
Kars	150 360	9	9	149 932	74 514
Kastamonu	244 426	20	20	242 413	84 514
Kayseri	1 452 458	17	17	1 452 156	471 974
Kırklareli	317 057	21	21	316 199	119 468
Kırşehir	206 016	10	10	205 563	82 979
Kocaeli	2 130 006	13	13	2 130 006	754 312

continued on following page

Table 5. Continued

	municipal population	municipalities(number)	Municipality waste services	Population	Amount of waste collected(amount)
Konya	2 330 024	32	32	2 317 499	919 745
Kütahya	457 263	28	28	457 226	154 751
Malatya	750 491	14	14	743 768	224 501
Manisa	1 475 353	18	18	1 475 353	586 890
Kahramanmaraş	1 134 105	12	11	1 094 838	356 435
Mardin	895 911	11	11	843 379	243 842
Muğla	1 081 867	14	14	1 077 021	735 310
Muş	246 389	23	22	244 066	84 023
Nevşehir	260 090	24	23	258 326	107 823
Niğde	307 358	29	29	302 849	113 969
Ordu	770 711	20	20	733 382	233 783
Rize	248 508	18	18	245 041	76 772
Sakarya	1 110 735	17	17	1 105 917	397 976
Samsun	1 382 376	18	18	1 328 341	414 427
Siirt	251 688	12	12	251 688	85 289
Sinop	142 489	9	9	142 108	67 944
Sivas	501 627	24	24	499 410	186 391
Tekirdağ	1 187 162	12	12	1 187 162	515 336
Tokat	496 112	43	40	486 322	177 321
Trabzon	822 270	19	19	816 367	225 668
Tunceli	59 455	9	9	59 362	27 030
Şanlıurfa	2 237 745	14	14	2 200 693	551 767
Uşak	297 501	11	11	297 460	119 438
Van	1 118 087	14	14	1 113 340	419 673
Yozgat	321 001	36	36	320 830	127 910
Zonguldak	436 950	25	25	435 740	162 082
Aksaray	364 389	22	22	363 162	102 917
Bayburt	58 622	5	5	58 622	19 942
Karaman	216 455	11	11	216 346	65 228
Kırıkkale	254 376	11	11	254 376	78 349
Batman	563 857	11	11	561 743	173 737
Şırnak	447 677	19	19	447 559	223 324
Bartın	115 878	8	8	115 741	73 336
Ardahan	42 448	7	7	42 448	27 066

continued on following page

Table 5. Continued

	municipal population	municipalities(number)	Municipality waste services	Population	Amount of waste collected(amount)
Iğdır	136 417	7	7	136 417	50 244
Yalova	275 487	14	14	273 682	123 717
Karabük	193 728	7	7	193 577	63 086
Kilis	123 603	4	4	123 542	72 659
Osmaniye	466 143	14	14	465 572	148 514
Düzce	297 950	10	10	297 638	120 127

Source: TURKSTAT, 2025e

Table 6 shows waste treatment indicators in Turkey giving the year of 2022 and 2024 as below.

Table 6. Waste Treatment Indicators

Indicators	**2022**	**2024**
Waste generated(total)	860 563 173	1 060 900 811
Hazardous waste(total)	23 794 881	37 759 154
Hazardous mineral waste(amount)	23 787 879	37 746 060
Total amount of non-H waste	836 768 292	1 023 141 657
Non-H mineral waste(amount)	836 687 187	1 023 038 273
Overburden(Amount)	834 254 003	1 020 416 206
Treatment methods(Amount)		
Recovered within establishment	2	4
Sold/sent to licenced waste treatment facilities	3 631 480	3 374 124
Used for reclamation purposes in mining sites	21 561 493	22 256 614
Sent to incineration plant	506	448
Backfilling	180 180 759	132 244 921
Sent to heaps for overburden/tailing dam/controlled landfill site	654 625 604	902 780 338

continued on following page

Table 6. Continued

Indicators	2022	2024
Dumping onto land	311 468	144 248
Others	251 862	100 114

Source: TURKSTAT, 2025f

Turkey is a developing country with a dense population (Yildirim and Bostanci, 2025), and waste management is crucial in combating environmental pollution. The most up-to-date waste management data from the TURKSTAT website has been compiled here to understand waste management in Turkey. Selected key indicators related to waste management in Turkey, based on open access and current data provided by TURKSTAT (2025), are presented in Tables 1, 2, 3, 4, 5, and 6. TURKSTAT presents waste management indicators in tabular form on its website from a wide range of areas across Turkey, including all municipalities, manufacturing facilities (with 50 or more employees), thermal power plants, and mining operations. According to 2024 data, a total of 120 million tons of waste was recorded. A total of 32.3 million tons of waste was collected in municipalities (TURKSTAT, 2025).

The Ministry of Environment, Urbanization and Climate Change has shared some statistics related to the zero waste movement. Accordingly, the recycling rate was 34.92% in 2023. Regarding the waste management and recycling practices initiated with the Zero Waste movement (TC Çevre, Şehircilik ve İklim Değişikliği Bakanlığı, (2024):

- Since June 2017, the number of buildings/campuses that have transitioned to a zero waste management system has reached approximately 185,000.
- There is approximately 21 million people trained on zero waste concept.
- Our recycling rate was reached to 34.92% in 2023.
- Key target is to increase our recycling rate to 60% by 2035.
- The collected waste has been recycled, contributing 256 billion Turkish Lira to our country's economy.
- 2.6 billion kWh of energy has been saved,

- 819 million m3 of water has been saved,
- 104 million m3 of landfill space has been saved,
- 5.9 million tons of greenhouse gas emissions have been prevented,
  498 million trees have been saved, and
- 127 million barrels of oil have been saved.

Karaosmanoğlu (2025) included Minister Murat Kurum's remarks regarding the Ministry of Environment, Urbanization and Climate Change's "2025-2035 National Waste Management Strategy and Plan." Accordingly, it was stated that waste management facilities in Türkiye will be adapted to geographical, demographic, and economic conditions by 2035, and that analyses have been conducted. The analyses revealed that waste types such as municipal, medical, construction, and recyclable waste were analyzed in light of the necessary economic, geographical, and demographic factors. Decisions have been made to expand zero-waste practices, and it appears that the necessary financing and investment needs must be planned (Karaosmanoğlu, 2025).

CONCLUSION

Climate change is a globally accepted reality, a global challenge. Environmental pollution and natural resource depletion have made the impacts of climate change more real. Preventing environmental degradation can slow climate change and ensure the long-term preservation of resources. Therefore, sustainable economic approaches, green economics, and circular economy models can be considered as new economic models that can save lives (Yildirim, 2025; Yıldırım, 2025; Yıldırım et al., 2024; Yıldırım and Yıldırım, 2025). At this point, waste management and recycling practices have become a growing new field within the circular economy. The focus of the circular economy is on reusing waste within resources. This reduces waste accumulation in the environment and ensures efficient use of resources.

This study examines key indicators in Turkey to provide a comprehensive picture of waste management. Sectoral analysis is crucial in waste management, and waste management strategies should be determined based on sectoral needs. In addition, municipalities play a key role in waste management (TURKSTAT, 2025).

The circular economy model plays a key role in achieving the European Green Deal's goal of achieving climate neutrality for countries by 2050, while also helping drive industrial innovation. According to the National Waste Management Action Plan, Turkey aims to recycle 60% of its municipal waste by 2035. The recovery rate reached 30.13% in 2022. Furthermore, the Regulation on the Regulated Landfill of Waste, published in the Official Gazette dated March 26, 2010, and numbered 27533, stipulated that at least 60% of the municipal waste collected by 2035 (Türkiye Cumhuriyeti Cumhurbaşkanlığı Strateji Ve Bütçe Başkanlığı, 2023).

EY Parthenon (2025) presented a significant report on waste management and recycling in Türkiye. The report made some important observations regarding Turkey's future. Growth is expected in waste management and recycling, and capacity is expected to increase in Türkiye. The implementation of renewable energy incentives in industrial zones is recommended, and the importance of digital and smart technologies in waste management is emphasized.

Based on the tables and information compiled in the study, the following recommendations are made:

- Waste management can be made more effective by implementing circular economy models at the micro level in densely populated cities.
- Appropriate city-focused waste management policies should be developed.
- Waste management facilities of appropriate quality and size should be adapted to meet the needs of each city.
- Waste and recycling should be considered in urban transformation.
- Technology and smart technology investments should be prioritized in waste management.
- Effective waste management facilities should be activated through smart transformation.

- Waste management should be prioritized to prevent environmental pollution and protect natural resources, and incentives should be increased in this direction.

REFERENCES

Awan, U., & Sroufe, R. (2022). Sustainability in the Circular Economy: Insights and Dynamics of Designing Circular Business Models. *Applied Sciences (Basel, Switzerland)*, *12*(3), 1521. DOI: 10.3390/app12031521

Bellezoni, R. A., Adeogun, A. P., Paes, M. X., & de Oliveira, J. A. P. (2022). Ayoola Paul Adeogun, Michel Xocaira Paes, José Antônio Puppim de Oliveira, (2022). Tackling climate change through circular economy in cities. *Journal of Cleaner Production*, *381*(Part 1), 135126. DOI: 10.1016/j.jclepro.2022.135126

Çevre, T. C. Şehircilik ve İklim Değişikliği Bakanlığı, (2024). Sıfır Atık ile geri kazanım oranı % 35'e ulaştı., https://cygm.csb.gov.tr/sifir-atik-ile-geri-kazanim-orani-35e-ulasti.-haber-286897

Chen, X. (2022). Machine learning approach for a circular economy with waste recycling in smart cities. *Energy Reports*, *8*, 3127–3140. DOI: 10.1016/j.egyr.2022.01.193

Christis, M., Athanassiadis, A., & Vercalsteren, A. (2019). Aristide Athanassiadis, An Vercalsteren, (2019). Implementation at a city level of circular economy strategies and climate change mitigation – the case of Brussels. *Journal of Cleaner Production*, *218*, 511–520. DOI: 10.1016/j.jclepro.2019.01.180

Ding, Z., Wang, X., & Patrick, X. W. (2023). Zou, (2023). Barriers and countermeasures of construction and demolition waste recycling enterprises under circular economy. *Journal of Cleaner Production*, *420*, 138235. DOI: 10.1016/j.jclepro.2023.138235

European Investment Bank, (2023). The EIB Circular Economy Guide: Supporting the circular transition – May 2023.

European Parliament. (2023). Circular economy: definition, importance and benefits, https://www.europarl.europa.eu/topics/en/article/20151201STO05603/circular-economy-definition-importance-and-benefits

Karaosmanoğlu, A. (2025). Türkiye'nin 10 yıllık "Ulusal Atık Yönetimi Stratejisi ve Planı" hazırlandı, https://www.aa.com.tr/tr/ekonomi/turkiyenin-10-yillik-ulusal-atik-yonetimi-stratejisi-ve-plani-hazirlandi/3719383

Khan, S., & Haleem, A. (2021). Investigation of circular economy practices in the context of emerging economies: a CoCoSo approach. International Journal of Sustainable Engineering, 14(3), 357–367. DOI: 10.1080/19397038.2020.1871442

Klein, N., Ramos, T. B., & Deutz, P. (2020). Circular Economy Practices and Strategies in Public Sector Organizations: An Integrative Review. *Sustainability (Basel)*, *12*(10), 4181. DOI: 10.3390/su12104181

Liu, J., Wu, P., Jiang, Y., & Wang, X. (2021). Explore potential barriers of applying circular economy in construction and demolition waste recycling. *Journal of Cleaner Production*, *326*, 129400. DOI: 10.1016/j.jclepro.2021.129400

Mandpe, A., Paliya, S., Gedam, V. V., Patel, S., Tyagi, L., & Kumar, S. (2022). Circular economy approach for sustainable solid waste management: A developing economy perspective. Waste Management & Research. *Waste Management & Research*, *41*(3), 499–511. DOI: 10.1177/0734242X221126718 PMID: 36264045

Mikhno, I., Ihnatenko, N., Cherniaiev, O., Vynogradnya, V., Atstaja, D., & Koval, V. (2023). Construction waste recycling in the circular economy model, IOP Conference Series: Earth and Environmental Science, Volume 1126, 2nd International Conference on Environmental Sustainability in Natural Resources Management 31/10/2022 - 01/11/2022 Riga, Latvia, IOP Conf. Ser.: Earth Environ. Sci. 1126 012003, DOI DOI: 10.1088/1755-1315/1126/1/012003

Möslinger, M., Ulpiani, G., & Vetters, N. (2023). Giulia Ulpiani, Nadja Vetters, (2023). Circular economy and waste management to empower a climate-neutral urban future. *Journal of Cleaner Production*, *421*, 138454. DOI: 10.1016/j.jclepro.2023.138454

Nixon, K. D., Schyns, Z. O. G., Luo, Y., Ierapetritou, M. G., Vlachos, D. G., Korley, L. S. T. J., & Epps, T. H.III. (2024). Analyses of circular solutions for advanced plastics waste recycling. *Nature Chemical Engineering*, *1*(10), 615–626. DOI: 10.1038/s44286-024-00121-6

OECD. (2020). *The Circular Economy in Cities and Regions: Synthesis Report, OECD Urban Studies*. OECD Publishing., DOI: 10.1787/10ac6ae4-

Parthenon, E. Y. (2025). Atık Yönetimi ve Geri Dönüşüm Sektör Analizi, https://www.ey.com/tr_tr/technical/ey-turkiye-yayinlar-raporlar/atik-yonetimi-geri-donusum-sektor-analizi

Puntillo, P. (2023). Circular economy business models: Towards achieving sustainable development goals in the waste management sector—Empirical evidence and theoretical implications. *Corporate Social Responsibility and Environmental Management*, *30*(2), 941–954. DOI: 10.1002/csr.2398

Ragossnig, A. M., & Schneider, D. R. (2019). Circular economy, recycling and end-of-waste. Waste Management & Research. *Waste Management & Research*, *37*(2), 109–111. DOI: 10.1177/0734242X19826776 PMID: 30803418

Rolewicz-Kalińska, A., Lelicińska-Serafin, K., & Manczarski, P. (2020). The Circular Economy and Organic Fraction of Municipal Solid Waste Recycling Strategies. *Energies*, *13*(17), 4366. DOI: 10.3390/en13174366

Romero-Perdomo, F., Carvajalino-Umaña, J. D., Moreno-Gallego, J. L., Ardila, N., & González-Curbelo, M. Á. (2022). Research Trends on Climate Change and Circular Economy from a Knowledge Mapping Perspective. *Sustainability (Basel)*, *14*(1), 521. DOI: 10.3390/su14010521

Sasmoko, Z., Zaman, K., Malik, M., Awan, U., Handayani, W., Jabor, M. K., & Asif, M. (2022). Environmental Effects of Bio-Waste Recycling on Industrial Circular Economy and Eco-Sustainability. *Recycling*, *7*(4), 60. DOI: 10.3390/recycling7040060

Schroeder, P., Anggraeni, K., & Weber, U. (2019). The Relevance of Circular Economy Practices to the Sustainable Development Goals. *Journal of Industrial Ecology*, *23*(1), 77–95. DOI: 10.1111/jiec.12732

Serrano, T., Aparcana, S., Bakhtiari, F., & Laurent, A. (2021). Contribution of circular economy strategies to climate change mitigation: Generic assessment methodology with focus on developing countries. *Journal of Industrial Ecology*, *25*(6), 1382–1397. DOI: 10.1111/jiec.13178

Suárez-Eiroa, B., Fernández, E., Méndez-Martínez, G., & Soto-Oñate, D. (2019). Emilio Fernández, Gonzalo Méndez-Martínez, David Soto-Oñate, (2019). Operational principles of circular economy for sustainable development: Linking theory and practice. *Journal of Cleaner Production*, *214*, 952–961. DOI: 10.1016/j.jclepro.2018.12.271

Türkiye, U. N. P. D. (2021). The Zero Waste Project receives UNDP Turkey's first Global Goals Action Award, https://www.undp.org/turkiye/press-releases/zero-waste-project-receives-undp-turkeys-first-global-goals-action-award

Türkiye Cumhuriyeti Cumhurbaşkanlığı Strateji Ve Bütçe Başkanlığı. (2023). Katı Atık Yönetimi, On İkinci Kalkınma Planı 2024-2028, Özel İhtisas Komisyonu Raporu, Ankara, 2023. https://www.sbb.gov.tr/wp-content/uploads/2025/08/Kati-Atik-Yonetimi-Ozel-Ihtisas-Komisyon-Raporu_01082025.pdf

TURKSTAT. (2025). Atık İstatistikleri, 2024, https://data.tuik.gov.tr/Bulten/Index?p=Atik-Istatistikleri-2024-54134

TURKSTAT. (2025a), Manufacturing industry waste management statistics, 2022, 2024 (Ton -Tonnes), Atık İstatistikleri, 2024, https://data.tuik.gov.tr/Bulten/Index?p=Atik-Istatistikleri-2024-54134

TURKSTAT. (2025b). Waste statistics of manufacturing industry subsectors, 2022, 2024 (Ton-Tonnes), Atık İstatistikleri, 2024, https://data.tuik.gov.tr/Bulten/Index?p=Atik-Istatistikleri-2024-54134

TURKSTAT. (2025c). Municipal sewage sludge statistics, 2022, 2024, Atık İstatistikleri, 2024, https://data.tuik.gov.tr/Bulten/Index?p=Atik-Istatistikleri-2024-54134

TURKSTAT. (2025d). Municipal waste management statistics, 2024, Atık İstatistikleri, 2024, https://data.tuik.gov.tr/Bulten/Index?p=Atik-Istatistikleri-2024-5413

TURKSTAT. (2025e). Municipal waste services statistics, 2024, Atık İstatistikleri, 2024, https://data.tuik.gov.tr/Bulten/Index?p=Atik-Istatistikleri-2024-54134

TURKSTAT. (2025f). Mining establishments waste statistics, 2022, 2024 (Ton -Tonnes), Atık İstatistikleri, 2024, https://data.tuik.gov.tr/Bulten/Index?p=Atik-Istatistikleri-2024-54134

UNPD. (2023). What is circular economy and why does it matter? https://climatepromise.undp.org/news-and-stories/what-is-circular-economy-and-how-it-helps-fight-climate-change

Van Caneghem, J., Van Acker, K., De Greef, J., Wauters, G., & Vandecasteele, C. (2019). Waste-to-energy is compatible and complementary with recycling in the circular economy. *Clean Technologies and Environmental Policy*, *21*(5), 925–939. DOI: 10.1007/s10098-019-01686-0

Vivas, M. A. U., Amórtegui, L. S., & Guerrero-García-Rojas, H. (2025). Understanding the characteristics of the green, circular and bioeconomy models within the context of sustainable development. *Energy, Sustainability and Society*, *15*(1), 27. DOI: 10.1186/s13705-025-00528-w

Yang, M., Chen, L., Wang, J., Msigwa, G., Osman, A. I., Fawzy, S., Rooney, D. W., & Yap, P.-S. (2023). Circular economy strategies for combating climate change and other environmental issues. *Environmental Chemistry Letters*, *21*(1), 55–80. DOI: 10.1007/s10311-022-01499-6

Yildirim, S. (2022). The Consumer Role for Sustainable Development: How Consumers Contribute Sustainable Development Goals. In I. Management Association (Ed.), Research Anthology on Measuring and Achieving Sustainable Development Goals (pp. 872-888). IGI Global Scientific Publishing. DOI: 10.4018/978-1-6684-3885-5.ch047

Yıldırım, S. (2025). Unveiling Circularity: Challenges and Barriers in Transitioning to Circular Practices. In Singh, P., Daga, S., Yadav, K., & Mishra, V. (Eds.), *Circular Economy and Environmental Resilience*. Springer., DOI: 10.1007/978-3-031-93091-1_8

Yildirim, S., & Bostanci, S. H. (2025). Informal Recycling Sector in the Developing World and Circular Economy: Case of Türkiye. In Borthakur, A., & Singh, P. (Eds.), *Circular Economy and Green Transition in the Global South*. Springer., DOI: 10.1007/978-3-031-90827-9_3

Yildirim, S., Demirtas, I., & Yıldırım, D. C. (2025). Circular Practices in the Fashion Industry: Opportunities and Threats. In Borthakur, A., & Singh, P. (Eds.), *Circular Economy and Green Transition in the Global South*. Springer., DOI: 10.1007/978-3-031-90827-9_8

Yıldırım, S., & Kantarcı, T. (2022). A Review on Sustainability Policies of Businesses: Recycling and Waste Reduction. *Journal of Recycling Economy & Sustainability Policy*, *1*(1), 1–9. https://respjournal.com/index.php/pub/article/view/1

Yıldırım, S., Sevik, N., Kandpal, V., & Yıldırım, D. C. (2024). The Role of Green Brands on Achieving 2030 Sustainable Development Goals (2030 SDGs). In Kankaew, K., Chaudhary, S., & Widtayakornbundit, S. (Eds.), *Contemporary Management and Global Leadership for Sustainability* (pp. 141–162). IGI Global Scientific Publishing., DOI: 10.4018/979-8-3693-1273-5.ch009

Yıldırım, S., & Yıldırım, D. Ç. (2025). Adapting Smart Waste Management and Recycling Practices Among the Circular Economy. In Kandpal, V., Gunasekaran, A., Jaswal, A., & Mukherjee, D. (Eds.), *Rethinking Resources. Approaches to Global Sustainability, Markets, and Governance*. Springer., DOI: 10.1007/978-981-96-9055-8_18

Zero Waste Foundation. (n.d). https://sifiratikvakfi.org/en

ADDITIONAL READING

Bostancı, S. H., & Yıldırım, S. (2021). Sustainable Communities vs. Climate Refugees: Two Opposite Results of Climate Change. In Popescu, C. (Ed.), *Handbook of Research on Novel Practices and Current Successes in Achieving the Sustainable Development Goals* (pp. 298–319). IGI Global., DOI: 10.4018/978-1-7998-8426-2.ch015

Bostancı, S. H., & Yıldırım, S. (2022). The Role of Municipalities in Achieving Water Security: The Case of Turkey. In Castanho, R. (Ed.), *Handbook of Research on Sustainable Development Goals, Climate Change, and Digitalization* (pp. 268–286). IGI Global., DOI: 10.4018/978-1-7998-8482-8.ch017

Sevik, N., & Yıldırım, S. (2023). Sustainable Enterprises Post COVID-19 Pandemic: The Rise of Sustainable Fashion Enterprises. In Castanho, R. (Ed.), *Handbook of Research on Current Advances and Challenges of Borderlands, Migration, and Geopolitics* (pp. 343–359). IGI Global., DOI: 10.4018/978-1-6684-7020-6.ch017

Yildirim, S. (2020). The consumer role for sustainable development: how consumers contribute sustainable development goals. In Chkoniya, V., Madsen, A. O., & Bukhrashvili, P. (Eds.), *Anthropological Approaches to Understanding Consumption Patterns and Consumer Behavior* (pp. 325–341). IGI Global. DOI: 10.4018/978-1-7998-3115-0.ch018

Yıldırım, S. (2021). Do green women influencers spur sustainable consumption patterns? Descriptive evidences from social media influencers. *Ecofeminism and Climate Change*, *2*(4), 198–210. DOI: 10.1108/EFCC-02-2021-0003

Yildirim, S. (2021). 3R Principle in Household Energy Consumption Patterns: A Case of Turkish Households. In Danish, M., & Senjyu, T. (Eds.), *Eco-Friendly Energy Processes and Technologies for Achieving Sustainable Development* (pp. 88–107). IGI Global. DOI: 10.4018/978-1-7998-4915-5.ch005

Yıldırım, S., & Bostancı, S. H. (2021). The efficiency of e-government portal management from a citizen perspective: Evidences from Turkey, *World Journal of Science. Technology and Sustainable Development*, *18*(3), 259–273. DOI: 10.1108/WJSTSD-04-2021-0049

Yıldırım, S., & Candan, B. (2015). Segmentation of green product buyers based on their personal values and consumption values. *Environmental Values*, *24*(5), 641–661. DOI: 10.3197/096327115X14384223590212

Yildirim, S., Demirtas, I., & Yildirim, D. C. (2022). A Review of Alternative Economic Approaches to Achieve Sustainable Development: The Rising Digitalization and Degrowth Post COVID-19. In Castanho, R. (Ed.), *Handbook of Research on Sustainable Development Goals, Climate Change, and Digitalization* (pp. 288–307). IGI Global. DOI: 10.4018/978-1-7998-8482-8.ch018

Yıldırım, S., & Kantarcı, T. (2022). A Review on Sustainability Policies of Businesses: Recycling and Waste Reduction. *Journal of Recycling Economy & Sustainability Policy*, *1*(1), 1–9. https://respjournal.com/index.php/pub/article/view/1

Yildirim, S., Kaplan, M., & Yildirim, D. C. (2023). Climate Change and the Ocean Economy: A Review on OECD's Ocean Economy. In Pego, A. (Ed.), *Handbook of Research on Bioeconomy and Economic Ecosystems* (pp. 212–232). IGI Global., DOI: 10.4018/978-1-6684-8879-9.ch012

Yıldırım, S., Sevik, N., Kandpal, V., & Yıldırım, D. C. (2024). The Role of Green Brands on Achieving 2030 Sustainable Development Goals (2030 SDGs). In Kankaew, K., Chaudhary, S., & Widtayakornbundit, S. (Eds.), *Contemporary Management and Global Leadership for Sustainability* (pp. 141–162). IGI Global., DOI: 10.4018/979-8-3693-1273-5.ch009

Yıldırım, S., & Vasilev, V. (2024). The Future of Green Collar Workers: Green Transition in Employment. In Yıldırım, S., Yıldırım, D., Demirtaş, I., & Kandpal, V. (Eds.), *Green Transition Impacts on the Economy, Society, and Environment* (pp. 1–16). IGI Global., DOI: 10.4018/979-8-3693-3985-5.ch001

Yıldırım, S., & Yıldırım, D. Ç. (2020). Achieving Sustainable Development Through a Green Economy Approach. In Patti, S., & Trizzino, G. (Eds.), *Advanced Integrated Approaches to Environmental Economics and Policy: Emerging Research and Opportunities* (pp. 1–22). IGI Global. DOI: 10.4018/978-1-5225-9562-5.ch001

Yıldırım, S., & Yıldırım, D. C. (2022). Achieving Seafood Security in the Mediterranean Region: A Case of Turkey. In I. Management Association (Ed.), Research Anthology on Ecosystem Conservation and Preserving Biodiversity (pp. 1780-1800). IGI Global. DOI: 10.4018/978-1-6684-5678-1.ch086

Yıldırım, S., Yıldırım, D. C., Demirtaş, I., & Kandpal, V. (Eds.). (2024). *Green Transition Impacts on the Economy, Society, and Environment*. IGI Global., DOI: 10.4018/979-8-3693-3985-5

Yildirim, S., Yildirim, D. C., & Erdogan, F. (2022). The Sustainability Level of Agricultural Food Markets: Selected Indicators From Mediterranean Countries. In Castanho, R. (Ed.), *Analyzing Sustainability in Peripheral, Ultra-Peripheral, and Low-Density Regions* (pp. 131–155). IGI Global., DOI: 10.4018/978-1-6684-4548-8.ch008

Yildirim, S., Yildirim, D. C., & Erdogan, S. (2024). Energy Security Dilemma in OECD Countries: The COVID-19 Pandemic, Russia-Ukraine War, and Energy Issues. In R. Castanho (Ed.), Green Economy and Renewable Energy Transitions for Sustainable Development (pp. 83-105). IGI Global. DOI: 10.4018/979-8-3693-1297-1.ch005

Yildirim, S., Yildirim, D. C., & Esen, Ö. (2023). Environmental Protection and Environmental Protection Expenditure in Developing Countries: A Case of Turkey. In Castanho, R. (Ed.), *Handbook of Research on Current Advances and Challenges of Borderlands, Migration, and Geopolitics* (pp. 100–119). IGI Global., DOI: 10.4018/978-1-6684-7020-6.ch005

KEY WORDS AND DEFINITIONS

Circular Economy: This is an economic model that, unlike a linear economy, involves a cycle in which waste is reintroduced into the system as a resource input.

Circular Practices: Sustainable processes within the circular economy.

Waste Management: Waste management is the management of the disposal, recycling, and recovery of unused materials that would otherwise accumulate in the environment as waste.

Chapter 5
Navigating Climate Action and Sustainability:
From Energy Justice to Waste Management in the Face of Climate Change

Smriti Tandon Gupta
https://orcid.org/0000-0003-4999-5881
Graphic Era University, India

Arvinder Kaur
https://orcid.org/0000-0001-7360-5674
Chandigarh University, Mohali, India

Pawan Kumar
https://orcid.org/0000-0003-4892-6374
Graphic Era University, India

ABSTRACT

Integrating energy justice with sustainable waste management, this study constructs a comprehensive framework addressing climate change by fostering socio-economic resilience, minimizing environmental

DOI: 10.4018/979-8-3373-9685-9.ch005

degradation, and promoting decentralized renewable energy systems through waste-to-energy technologies; employing an interdisciplinary methodology—encompassing statistical modeling, geospatial analysis, global case studies, and stakeholder interviews—the research identifies systemic barriers such as financial constraints, technological limitations, and policy fragmentation while advancing circular economy principles and participatory governance models to harmonize sustainability with social equity, ensuring transformative, inclusive, and equitable global climate resilience strategies.

INTRODUCTION

These urgent threats establish climate change as an essential case for complex solutions beyond conventional environmental and technological frameworks. This study explores the convergence between sustainable waste management and energy justice while investigating their role as fundamental strategies for combating climate change and advancing societies' transition into new architectural networks which balance environmental sustainability with socioeconomic goals (Popa et al., 2010). The research employs equity concepts and Innovation and circular economy frameworks to show how renewable energy systems and fair resource distribution work toward cutting climate change impacts on vulnerable demographic segments. This paper explores the development of an analytical framework for climate justice to integrate climate change programs with social equity goals towards building a transformative global system (Yuan et al., 2024).

Several environmental indicators highlight a worrying pattern requiring a review of production and consumption behavior. Human use of resources without mindful preservation and inadequate waste disposal resulting from indifferent environmental priorities have caused climate change, repeated global warming, weather pattern anomalies, and biodiversity decline (Destrée, 2024). Integrating energy justice into climate transformation procedures becomes vital to eliminate the enduring social injustices that maintain environmental and socioeconomic marginalization. Energy justice is a heuristic toolkit that prioritizes appropriate energy resource allocation and tackles power poverty while enhancing the quality of energy delivery systems for communities experiencing direct climate change effects. Energy justice operates on three foundational tenets: Researchers divided energy justice into four fundamental categories: distributive justice and procedural justice, together with recognition justice and interactional justice (Ahlborg et al., 2024).

We ensure equitable distribution of energy profits and losses among social classes through distributive justice mechanisms to tackle unfair environmental costs and energy access limitations. Implementing procedural justice supports fair decision-making methods, which enhance everyone's power to feel the impact of energy choices (Fang et al., 2023). Following justice recognition, one reveals how specific groups experience social exclusion when positioning their rights as vital participants in energy policy decision-making. The foundation of global circular economy implementation depends on sustainable waste management practices as an indispensable support mechanism. The current model of utilities leads to natural capital erosion by removing resources and releasing waste from their material use to disposal operations. The circular economy builds lasting economic growth by designing circular material processes that boost resource productivity and dynamically transform waste streams into useful products. The new paradigm requires Waste-to-energy (WTE) technologies to achieve its dual purpose of renewable energy generation combined with landfill reduction (Trakic et al., 2024). Several WTE conversion technologies that perform, such as anaerobic digestion gas, gasification and pyrolysis, serve as platforms to transform non-recycled organic waste into usable energy products with reduced greenhouse gas generation rates and better national security protection. Achieving energy justice in sustainable waste management demands implementing multiple technical components while ensuring social benefits, ecological sustainability, and broader technological developments. Lines of innovative development are essential, but they do not generate appropriate solutions for climate mitigation and adaptation. Public policy frameworks and sound innovation practices should guide distribution allocations of these advancements. Community involvement and governance practises represent essential socio-cultural elements determining climate intervention success rates. In partnership with policy frameworks and socio-cultural processes, technology is the foundation for systemic change methods that deliver effective solutions that promote fairness.

Data shows that marginal groups who experience physical exposure to climate shocks and suffer resource degradation face the greatest impacts (Zhong et al., 2024). These groups include people living in developing nations, First Nations communities, and SIDS populations. The communities face worse conditions because they receive insufficient and expensive energy services from their providers. The energy divide requires decentralized renewable electricity systems with microgrids and community-shared solar programs to provide sustainable electric access while improving power availability and security and reducing dependence on traditional fuel sources. They also do more than merely lessen energy poverty: Renewable power solutions assist adaptation and mitigation efforts regarding climate change. The approach of energy justice exists alongside energy democracy, which focuses on creating decentralized democratic energy systems. Community self-governance over energy assets and local-level energy systems self-organization define energy democracy (Njangang et al., 2024). Through human capital development, workforce creation, and socioeconomic progress, the adoption of energy democracy strengthens local communities' security while raising their economic empowerment levels. Energy democracy allows the development of better climate strategies that shift existing top-down-market models to decentralized people-centered approaches for climate justice.

Circular economy standards recognize waste management as a core principle emphasizing that waste decomposition should lead to a future where waste functions as valuable materials, not discarded residues (Ang et al., 2024). Global environmental impacts are significantly reduced when waste streams undergo reprocessing through recycling, material recovery, and upcycling methods for resource conservation. Building WTE into waste management systems presents an opportunity to address two pressing issues: Solid waste storage requirements decrease simultaneously with the implementation of new power generation facilities. Renewable biogas energy made through organic waste anaerobic digestion processes is an energy source for electrical needs and vehicle and heating applications. Pea waste conversion into synthetic fuels and chemical feedstock generates next-generation energy supplies to boost energy diversification across industries and advance their maturity (Marshall & Pearse, 2025).

This framework establishes vital regulatory frameworks to ensure energy justice principles interact constantly with sustainable waste management systems. Strategies for policy implementation need supporting instruments to utilize regulatory frameworks through economic stimulants and technical solutions for building renewable systems and waste-to-energy protocols. The combination of feed-in tariffs, renewable energy credits, and tax incentives function as investment attractors for constructing distributed networks of energy systems along with waste management improvements. Enterprises face major operational obstacles, which combine with environmental laws and waste reduction requirements to establish environmentally sustainable processes for industry and municipalities. Before making policy decisions, judicial bodies must gather stakeholders at multiple meetings to recognize where interest groups can find common ground. Climate solutions that embrace principled equity become vital because scientific research shows how underserved communities suffer disproportionately from climate change impacts. Climate change and its related events force vulnerable social groups to relocate and result in insufficient access to food and water while generating poor health conditions, thereby expanding societal disparities, according to Marami et al. (2024). The scientific consensus shows that successful climate change solutions must use individual strategies that reduce poverty levels while expanding women's empowerment and civil rights. A gender-friendly approach within energy systems and waste management creates circumstances for women to emerge as influential agents in climate change dialogues, which yields better climate change outcomes (H. Kaur et al., 2023). When researchers apply energy justice frameworks and sustainable waste management concepts, they gain a more comprehensive grasp of climate change. The research identifies the current acceptance of technological implementation in social structures and policy frameworks before introducing various system-wide solutions for environmental well-being and social justice. This research will contribute new insights to climate justice understanding that will lead to interdisciplinary solutions for global sustainable development goals. This study aims to advance the cultural development goal through tolerance development hum, adaptation and a swift societal transition toward sustainability (A. Kaur & Sharma, 2024).

Table 1. Key Climate Indicators and Targets for 2030

Key Climate Indicators	Current Levels	Target Levels (2030)
Global CO2 Emissions	36.3 Gt/year	25.0 Gt/year
Renewable Energy Share	29%	50%
Waste Recycled	19%	50%

Source: International Energy Agency (2023).

Table 1 presents current levels and 2030 targets for key climate indicators, reflecting the global goals for reducing CO2 emissions, increasing the share of renewable energy, and improving waste recycling rates

THE NEXUS OF ENERGY JUSTICE AND CLIMATE ACTION

Defining Energy Justice

Every citizen should enjoy equal access to energy resources from which disadvantaged groups benefit most. The approach advocates fair distribution of energy production consumption costs and regulatory values. The evaluation must combine analysis of energy systems' social effects on marginalized communities and their impacts on both natural environments and economic outcomes for less privileged social strata (Dong et al., 2024). Under energy justice principles, communities must participate in energy-planning decisions to actively transform their adversities into actionable solutions. The framework operates on three core tenets: Three categories of justice combine distributive approaches to resource allocation with procedural decision-making fairness and recognition methods to address past and current community exclusion. These principles work together to create a new social and power arrangement from which accessibility to fair electricity services can be extended to all people, particularly to societies facing marginalization.

Table 2. Key Principles of Energy Justice

Principle	Definition
Distributive Justice	Equitable allocation of energy resources and benefits (Fang et al., 2023; Heffron, 2024).
Procedural Justice	Inclusive participation in energy decision-making processes (Bhullar, 2023).
Recognition Justice	Acknowledgment of historically marginalized groups in energy policies (Heffron & Sokołowski, 2024; Min & Lee, 2024).

Source: Energy Justice Framework (2023).

Table 2 outlines the three core principles of energy justice, which focus on ensuring fairness in resource distribution, inclusivity in decision-making, and the recognition of marginalized communities in energy policy.

Barriers to Energy Justice

While world countries are implementing changes to adopt renewable energy sources, some systemic challenges persist and act as barriers: unfair economics, infrastructure, and governance. Renewable energy adoption continues to severely limit underprivileged rural populations who often fail to join renewable energy shifts. People in marginalized communities face energy discrimination because they lack affordable and strong energy service options (Haldar et al., 2023). These communities face difficulties because their present energy requirements demand new infrastructure systems. The final energy management policies and organizational implementation deficiencies create worse energy problems because vulnerable populations lack fair representation. Worldwide efforts to adopt renewable energy suffer significant setbacks from energy insecurities, simultaneously making communities more vulnerable to climate change effects while seeking renewable transitions. These groups experience dire conditions because they simultaneously combat climate change despite the ongoing deterioration of their ecosystems from extreme weather phenomena caused by changing climates. These populations lack access to alternate coping mechanisms. The regions are caught in a vicious cycle linking energy difficulties with climate sensitivities that produce enduring discrimination. This expands Mother Nature's challenges, which diminishes access to needed resources. Renewable energy transformation faces a problem since it does not solve systemic barriers, thus potentially deepening societal inequality by establishing a new global standard dominated by patriarchal capitalist white middle-aged male norms that enable unfairness (Trakic et al., 2024).

Recent Study: According to an International Energy Agency 2023 report, 789 million people worldwide lack electricity access, predominantly in Sub-Saharan Africa.

STRATEGIES FOR ENERGY EQUITY

To bridge the existing gaps in energy access and equity, initiatives must focus on the following key strategies:

Waste Management: A Cornerstone of Sustainability

- Expanding Renewable Energy Infrastructure in Underserved Regions: Remote populations with disadvantaged communities require access to programs that deliver renewable home power generation systems using solar, wind turbine and hydroelectric technologies. Better access to energy supply exists alongside poverty reduction and decreased fossil fuel dependence after expanding infrastructure systems (Babalola & Olawuyi, 2022).
- Subsidizing Clean Energy Technologies for Low-Income Households: Through financial aid programs such as subsidy grants and affordable loans, low-income families can purchase clean energy technology components, including solar panels, along with energy-saving appliances and home insulation devices to decrease their energy expenses and environmental impact (Sovacool et al., 2022).
- Fostering Participatory Governance Models: When communities actively participate in energy decision-making, they help create policies that specifically serve the needs of their population (Ramasar et al., 2022). Through participatory governance, communities gain representation and influence in developing energy policies that determine the opportunities and welfare of their residents. Proactive

approaches build equity in the future energy system by providing sustainable, clean power to persons from all economic backgrounds.

The Escalating Waste Crisis

The growing speed of urbanization, rising consumption levels, and expanding populations worldwide continue to create environmental and health dangers from global waste production. Waste production exceeds acceptable levels while cities grow, and large waste management systems globally become overwhelmed. Unmanaged waste streams that result in disorderly open dump sites generate two main environmental issues by allowing organic waste and plastic and e-waste build-up (Kaswan, 2023). Healthy ecosystems and species stability become jeopardized when publisher management fails to keep pollutants away from land and water bodies or air. Rate this introduction: Concerning health outcomes that result from how we treat waste while inefficient environmental effects mark our present disposal approaches. Skin illnesses and infections result from hazardous waste and polluted water and soil, while these elements generate respiratory diseases, cancer, waterborne disease, and neurological disorders. Individuals from low-income areas and underserved communities face the most significant exposure to environmental hazards because they live close to waste sites or have inadequate waste disposal infrastructure. The continuing poverty and health gap undermines development efforts in overall health and equilibrium maintenance efforts. Waste management strategies alone cannot tackle the worldwide waste crisis. Public health policy collaborations with waste efficiency work toward reducing waste residues while promoting responsible consumption to protect human health (Adhikari et al., 2024).

Recent Study: According to UN data from 2022, the global solid waste output stands at 2 billion tons per year, but recycling levels reach just 13.5%.

Integrated Waste Management Systems

The successful implementation of integrated waste management needs multiple approaches activated throughout the waste management phases, from production to disposal. Physical waste management demands implementation because it protects nature while preserving materials and advancing sustainability efforts. Key components of this integrated strategy include:

- **Source Reduction**: Sustainability focuses on developing practices that help customers use fewer resources and minimize production waste (Biakhmetov et al., 2024). Industry waste generation can be substantially decreased when products extend their life cycle, require less material, and avoid superfluous packaging. Such proactively managed operations save natural resources while producing less waste, further decreasing industrial environmental impact throughout production and consumer use.
- **Circular Economy Models**: Under the degrowth model's circular economy framework, products and materials must be designed for maximum possible reuse, refurbishment and recycling before being landfill-bound (Jeetah et al., 2024). The strategy exists to generate valuable resources from waste management or resource recovery. Product recycling and recovery alongside reuse help extend the ownership duration in closed economic cycles that decrease raw material use and energy needs and minimize the greenhouse effect.
- **Technological Innovations**: The modern waste management sector transforms its core practices with newly developed AI and IoT combination technologies. Modern waste management operations benefit from linking collection procedures and waste materials with these modern technologies. AI enables waste recycling automation and precise sorting operations that boost performance results (Anokye et al., 2024). Implementing IoT sensors facilitates waste-level observation while optimizing collection routes and delivering immediate data for enhanced decision-making. Technology improvements decrease operational problems while decreasing financial expenses to improve the overall waste management system performance. These strategies implemented in waste management systems will create a resource-efficient world while achieving waste reduction, natural resource conservation, and extensive environmental and health benefits.

Table 3. Impact of Waste Management Strategies on Emissions Reduction

Waste Management Strategies	Impact on Emissions Reduction
Recycling	40% reduction in landfill waste
Composting	30% reduction in organic waste
Waste-to-Energy	25% reduction in CO2 emissions

Source: United Nations (2022).

Table 3 illustrates the significant contribution of various waste management strategies to emissions reduction, emphasizing the environmental benefits of recycling, composting, and waste-to-energy practices.

Policy and Community Engagement

Sound policy instruments and an engaged community are imperatives for effective and sustainable waste management. Waste generation needs to be tightly controlled by reforms; recycling requirements have to be complied with, and circular economy practices need to be promoted to cut off the use of landfills. Policies should also facilitate the introduction of advanced technologies in the processing of waste and recovery of resources. At the same time, community-based approaches effectively support extreme bottom-up waste reduction processes and enhance responsibility (Huang, 2024). Partnerships between governments, NGOs, and communities can also strengthen awareness raising, support, and action in consuming sustainably and anchoring local practices into national and international waste management frameworks. The employment of this approach is considered key to attaining a sustainable environment and fair disposal and management of waste for social benefits.

INTERDISCIPLINARY APPROACHES TO CLIMATE RESILIENCE

Synergizing Energy and Waste Management

The intersection of energy and waste systems presents significant opportunities for innovation, offering novel solutions to energy generation and waste management. Sustainable and circular systems emerge through the fusion of tourism and hydropower management sectors to optimize resource utilization and reduce environmental impact. Some key opportunities for innovation include:

- **Utilizing Waste-to-Energy Technologies**: Waste-to-energy (WTE) systems will help future waste management technology convert organic waste into useful energy resources. The industrial application of anaerobic digestion techniques converts organic waste into biogas products that generate electrical power and thermal energy, as described by Sumiyati et al. (2024). Waste conversion creates valuable energy resources from disposed waste while decreasing landfill size until communities can utilize power generation sites. Waste-to-energy systems can produce energy while stopping greenhouse gas production by moving waste materials from traditional disposal sites to new energy generation systems.
- **Developing Decentralized Energy Grids Powered by Renewable Waste Streams**: Creative development includes decentralized energy networks that obtain power from renewable waste streams as an energy supply. Without depending on central electricity supply systems, these decentralized waste management systems can still make renewable electricity from local waste types, including agricultural residue, food waste, or industrial waste by-products. Such systems are preferred in rural areas or regions with limited connection to the primary electricity distribution grid (Ghamari et al., 2024). This distributed generation also benefits energy security, reduces energy transmission losses, and provides for proper waste disposal as waste becomes an energy source. In addition, decentralized energy grids may be connected with innovative technologies aimed at improving

the organization of energy distribution and usage, increasing general efficiency, and, therefore, ensuring the sustainability of energy grids. What is more important is to find the symbiosis of the energy and waste systems and turn waste into efficient energy sources and a circular economy. Their applicability assists in meeting critical environmental concerns, minimizing reliance on fossil resources, and advancing superior waste and energy methods.

Policy Synergies

Governments must achieve policy coherence where energy justice and waste management form part of coherent climate policies since the two sectors are very connected in their environmental, social, and economic interventions (Patil et al., 2024). Such a mechanism ensures that energy and waste policies work together to enhance sustainability solutions by embracing a social justice dimension to the overall goals such as reduction of greenhouse emissions, social equity, and constructiveness resource use. Several policy measures can be implemented to achieve these objectives:

- **Carbon Pricing Mechanisms**: Governments can adopt carbon pricing mechanisms, such as carbon taxes or cap-and-trade systems, to create financial incentives for reducing carbon emissions. Revenues generated from these mechanisms could be earmarked to fund renewable energy projects, energy efficiency initiatives, and waste reduction programs. Governments can simultaneously address the dual challenges of climate change and waste disposal by directing these funds toward renewable energy development and innovative waste management solutions (Ezz et al., 2024). These funds could also support energy access for marginalized communities, ensuring that the transition to clean energy is equitable and inclusive.
- **Integrated Renewable Energy and Waste Reduction Programs**: Policy frameworks that link renewable energy development with waste reduction efforts can help create a circular economy that reduces reliance on fossil fuels while minimizing waste. For instance, governments can incentivize waste-to-energy technologies or support creating decentralized energy systems powered by renewable

waste streams (Aromolaran & Sartaj, 2024). Additionally, policies could promote the establishment of sustainable waste management practices, such as recycling, composting, and waste diversion from landfills, alongside the development of clean energy infrastructure. By promoting these synergies, governments can contribute to the dual goals of reducing carbon emissions and waste while simultaneously creating green jobs and stimulating local economies.

- **Inclusive Stakeholder Engagement**: Energy justice principles should be incorporated into waste management policies to ensure marginalized communities actively engage in decision-making processes. Governments can implement participatory mechanisms that allow local communities, particularly those most affected by environmental injustices, to have a say in energy and waste management policies. This would ensure that the benefits of clean energy and waste reduction programs are equitably distributed and that vulnerable populations are not left behind in the transition to a sustainable future.

By adopting these harmonized policies, governments can accelerate climate action while fostering a more inclusive and resilient society. This integrated approach tackles pressing environmental challenges and ensures that the transition to a low-carbon, circular economy benefits all sectors of society, particularly those most vulnerable to climate change and ecological degradation.

Community and Global Partnerships

Collaborative efforts between governments, non-governmental organizations (NGOs), and international organizations are essential for tackling climate change, energy justice, and waste management (Ram & Bracci, 2024). While global initiatives like the Paris Agreement provide a vital framework for collective action, they must be complemented by localized strategies tailored to different regions and communities' specific needs and contexts.

Global agreements set ambitious targets and foster international cooperation. Still, they must be supported by on-the-ground actions that address local realities, such as socioeconomic conditions, cultural practices, and environmental challenges (Banaś et al., 2024). This requires a collaborative approach, where various stakeholders work together to design and implement context-sensitive solutions that reflect local communities' unique needs and priorities.

- **Governments** must play a central role by enacting policies aligning with international climate goals while considering their regions' socioeconomic conditions and environmental priorities. Local governments, in particular, can leverage their proximity to affected communities to craft tailored policies that address climate action's global and local dimensions (Cho et al., 2024).
- **NGOs** bring essential expertise, community engagement, and grassroots connections, which are crucial for identifying the most pressing needs and ensuring that solutions are inclusive and equitable. By working alongside communities, NGOs can help bridge gaps between policy and practice, empowering local populations to participate in climate action and waste management efforts.
- **International Organizations** provide vital technical support, financial resources, and knowledge-sharing platforms. They help coordinate global efforts and ensure that initiatives are scalable and adaptable to different regions. They also play a key role in fostering dialogue between governments, NGOs, and other stakeholders to promote a shared vision of sustainable development and climate justice (Paxton et al., 2023).

In addition to global frameworks, **localized actions** are needed to ensure that climate solutions are effective, inclusive, and sensitive to local contexts. These actions can include promoting renewable energy solutions that fit the needs of specific communities, implementing waste management systems tailored to local capabilities, and ensuring that marginalized groups are included in decision-making processes. By combining global commitment with localized strategies, we can achieve more comprehensive, resilient, and equitable climate solutions that leave no one behind.

CASE STUDIES AND BEST PRACTICES

Scandinavian Innovations

Sweden and Denmark's countries appropriately illustrate the functionalities of WTE systems and how variation and quality innovations in waste management can help decrease the dependency on landfills and produce renewable energy simultaneously. The two countries have achieved efficient methods for managing waste by applying advanced WTE technologies, which in turn assist the nations in meeting their power needs. This implies that circular economy models could solve environmental and energy problems (Ibarra-Esparza et al., 2023). WTE systems are applied in Sweden as a critical component of the country's waste management system. The Swedish waste-to-energy system is well-developed since the country uses waste to produce heat and electric power for thousands of homes and industries. The accomplishment of this strategy has enabled Sweden to rank among the countries with the highest recycling and waste diversion rates, if not the highest, and considerably decrease the quantity of waste that is disposed of in a landfill. However, Sweden currently disposes of only a tiny percentage of its waste in landfills and transforms the greatest amount of waste into energy, of which a great portion is renewable. The district heating systems across the country, with WTE plants among the main supply sources, remain a key supplier of urban heating and support Sweden's breakthrough climate targets.

Likewise, Denmark has a long way to go in the energy mix by integrating waste-to-energy systems. The Danish case also focuses on the function of WTE in Denmark's plan to tackle Carbon emissions and sustainable waste management. Denmark already has a policy whereby WTE plants assist in reducing landfill usage while producing electricity and district heating from the waste (Shah et al., 2024). Denmark has been developing new approaches, including where energy sourced from waste is used to generate both electric energy and useful heat, with the resultant heating produced by large district heating networks encompassing most major cities. This integrated system management services waste disposal and energy consumption and contributes to Denmark's environmental efficiency and sustainable energy. As seen in Sweden and Denmark, energy waste can be a force for cutting emissions, managing waste, and creating renewable energy. Their success further underscores the benefit of disruptive, circular business models that transform waste from a nuisance to a productive asset and align with national and global sustainability agendas. These countries' experiences can serve as important examples for other countries and their attempts to implement similar systems that would help achieve energy security, decrease greenhouse gas emissions, and support the goal of transitioning to a circular economy (Wilson, 2023).

Table 4. Landfill Rates and Energy Recovery from Waste in Sweden and Denmark

Country	Landfill Rate (%)	Energy Recovered from Waste (%)
Sweden	1%	50%
Denmark	2%	47%

Source: Scandinavian Waste-to-Energy Studies (2023).

Table 4 highlights these countries' highly efficient waste management systems, achieving very low landfill rates and a significant portion of energy recovery from waste. Sweden, for example, has reduced landfill usage to just 1% of its waste, with half of the waste being converted into energy. Denmark also demonstrates a highly effective waste-to-energy system, recovering 47% of its waste to generate renewable energy. These figures underscore the successful integration of waste management and energy recovery strategies in these nations, positioning them as global leaders in sustainable waste management and renewable energy generation.

Community-Led Energy Initiatives in Sub-Saharan Africa

This study provides a detailed view of how grassroots solar projects in Kenya show the potential of local investment and access technologies in filling the energy gaps, especially in developing and rural regions. Such efforts demonstrate the potential of solar electricity to bring light in areas where no grid exists or is wanting, in as much as it empowers households and societies to control their fate by generating electricity with solar panels that withstand test, time, and space on the one hand, and economic and ecological adversity on the other. This has revealed that while only a small section of Kenya is connected to the national electricity grid, hence the low legal standard of access to solar power in many Kenyan rural areas, decentralized grassroots solar power projects serve as life support in many parts of the country. Small-scale solar installation has been pioneered by local organizations, NGOs, and other community-based organizations and institutions to deliver bespoke solar solutions for homes, schools, health centres, or small businesses. These solar activities entail the provision of cheap photovoltaic modules of solar home systems, solar lamps, or solar water pumps, which serve as a clean source of reliable HH lighting instead of kerosene, which is costly, toxic, and polluting the users and the environment.

What distinguishes these projects is increased attention to local actors' participation in the activity's development and implementation. Most schemes entail community involvement, so the solutions implemented are culturally suitable, inclusive, and derived from the community lakes. These solar projects? Due to collaboration with the locals on how the solar projects are powered, executed, and maintained, the community becomes empowered and thus able to meet its energy needs independently. Also, developing local personnel to install and maintain solar products offers business opportunities while at the same time building the capacity of the community (Perkumienė et al., 2023). They also explain how these grassroots solar initiatives positively impact the stability of vulnerable communities by increasing reliance on fossil fuels and increasing access to energy for many important and necessary services. For example, solar health clinics can continue running without power outages in the middle of their working hours. Likewise, photonic power can be used to pump water, reminiscent of the assembly of electrical pumps that bring water closer to people to reduce the extensive journeys made by women and children to distant sources and bring water home. At the same time, the flexibility and cost-effectiveness of solar power systems enable their implementation in relatively isolated regions where it is impossible to construct energy networks. As costs for solar assets continue to reduce, these generation projects are expanding the coverage of easy access to electricity to increased numbers of low-income clients, hence promoting the achievement of universal energy access and SDGs. Therefore, it can be concluded that using simple, low-cost solar-emerging projects in Kenya shows that the issue of access to energy coupled with sustainable development is possible, even in volatile contexts, by applying appropriate innovative technologies complemented by a focus on community-based approaches (Saeed et al., 2024). These initiatives give people in other areas of the globe to experience similar problems with a scalable, swiftly applicable model of change that can help speed up the transformation towards the use of renewable energy sources and concurrently address so many key social, economic, and environmental wants.

CHALLENGES AND FUTURE DIRECTIONS

As the world moves toward more sustainable and resilient systems, several challenges must be addressed to realize the full potential of climate solutions. These challenges include overcoming technological and financial barriers, navigating policy fragmentation, and prioritizing equity in climate action. The following sections outline these challenges and suggest future directions for tackling them (Nawar et al., 2024).

Overcoming Technological and Financial Barriers

The widespread adoption of innovative climate solutions remains restricted by expensive equipment and complicated technological complexity. Energy-efficient infrastructure, renewable energy systems, and waste-to-energy technologies present installation hurdles to low-income groups, small companies, and developing nations. Some regions face challenges because they lack the essential technological knowledge and supporting infrastructure for these innovations to grow at scale.

Research initiatives combined with capacity-building efforts must receive funding to develop technologies that become affordable while achieving both efficiency and local adaptability. Area-specific innovations require developing affordable modular components and implementing solutions that match particular community requirements. Without adequate financial mechanisms such as grants, interest loan subsidies, and climate financing, the democratization of sustainable technologies remains unrealized. Through collaborations between governmental entities, private entities, and international organizations, we can address financing gaps so climate solutions reach every segment of our population (Rossetto, 2023).

Navigating Policy Fragmentation

Effective climate action faces a major obstacle because different agencies use separate policies between regions and sectors. The absence of integrated policies results in fragmented climate change, energy access, and waste management activities that deliver reduced effectiveness. Nonaligned angles between energy policies, waste management, and environmental protection regulations create both efficiency losses and missed potential synergies in the system. National and regional variations in policy frameworks create barriers to solution scaling since cross-border cooperation becomes hampered for climate migration, cleantech trade, and the management of shared natural assets (Pillay et al., 2024; Singh et al., 2023).

The solution demands integrated governance structures that align policies from the energy, waste management, agriculture, and climate adaptation sectors. Stakeholders from multiple governance areas should form integrated climate action plans to unite sectors for coordinated strategy development and implementation. Clear regulatory frameworks must promote coherence between jurisdictions and facilitate successful cross-jurisdictional collaboration for climate initiatives.

Prioritizing Equity in Climate Solutions

Social equity should be an essential focus for the ongoing climate change response. Climate solutions must face environmental effects and proactively address historical and current social inequalities (Longfor et al., 2023). Vulnerable communities, which include lower-income and rural people alongside marginalized groups, continue to face larger burdens from climate change effects that result in extreme weather serious health effects and forced displacement. Members of these populations face multiple disadvantages because they lack basic service access such as power grids, water supply systems, and healthcare facilities, so they suffer more from climate change impacts.

Regular climate strategies of the future need to advance social justice by correcting unequal access to environmental benefits while also providing marginalized groups the power to make decisions about climate action (Yap et al., 2024). They need control in creating climate plans and inclusive access to network connections, along with financial backing and adaptive technology to manage climate stress. The sustainability of a future direction depends on establishing strong governing systems that include every community and ensuring equal treatment through fair policy adoption because this process requires mandatory comprehensive community empowerment.

- **Managerial and social implications of the study:** The managerial implications of this study emphasize the need for cross-sectoral collaboration and strategic investments in renewable energy infrastructure and sustainable waste management technologies, urging leaders to prioritize equity-driven policies that enhance resource access and mitigate environmental inequalities while fostering innovation and capacity-building within marginalized communities (Tisi et al., 2023). Socially, the findings underscore the importance of inclusivity in climate action, advocating for the active participation of underrepresented groups in decision-making processes, facilitating grassroots empowerment, and addressing historical inequities through holistic, community-centred approaches, which collectively contribute to socio-economic resilience, environmental justice, and the advancement of global sustainability goals amidst escalating climate crises (Ibrahim et al., 2023; Karak et al., 2024).

CONCLUSION

Navigating the complexities of climate action requires a holistic approach incorporating energy justice and waste management into the broader sustainability agenda (Hai et al., 2023; Varjani et al., 2024). The challenges of climate change, environmental degradation, and social inequities are deeply interconnected, and solutions must be designed to address these issues comprehensively. By integrating sustainable energy practices, reducing waste, and ensuring that the benefits of these efforts reach marginalized and vulnerable communities, societies can build more resilient systems that not only mitigate the impacts of climate change but also promote long-term social and environmental well-being.

To achieve this, it is essential to foster innovation, equity, and collaboration across all sectors of society (Tereshchenko et al., 2023; Zhou & Zhang, 2022). Innovation in technology, finance, and policy will be crucial to driving the sustainable solutions needed to meet the pressing challenges of climate change. At the same time, ensuring that these innovations are inclusive and accessible will help ensure that the most vulnerable populations are not left behind. Moreover, collaboration—both at the local and global levels—is key to creating synergies that can accelerate the transition to a sustainable future.

The path forward demands an unwavering commitment to inclusivity, adaptability, and transformative action (Coskuner et al., 2021; Taşkın & Demir, 2020). As we develop solutions that benefit all people and the planet, the voices of marginalized communities must be heard and integrated into decision-making processes. Only through collective, coordinated, and inclusive efforts can we build a more just, sustainable, and resilient world that addresses the challenges of climate change and improves the quality of life for all people, now and in the future.

REFERENCES

Adhikari, S., Mahmud, M. A. P., Moon, E., & Timms, W. (2024). Comprehensive life cycle assessment of garden organic waste valorisation: A case study in regional Australia. *Journal of Cleaner Production*, *472*, 143496. Advance online publication. DOI: 10.1016/j.jclepro.2024.143496

Ahlborg, H., Michael, K., Unsworth, S. J., Hategekimana, S., Osunmuyiwa, O., Åberg, A., & Hultman, M. (2024). Thirty-five years of research on energy and power: A landscape analysis. *Renewable & Sustainable Energy Reviews*, *199*, 114542. Advance online publication. DOI: 10.1016/j.rser.2024.114542

Ang, P., Goh, W., Bu, J., & Cheng, S. (2024). Comparative Industrial-Scale Life Cycle Assessment of Base Case and Heat Recovery Scenarios for Carbon Capture from Natural Gas Combined Cycle Power Plants Using Aqueous Ammonia. *ACS Sustainable Chemistry & Engineering*, *12*(51), 18335–18349. DOI: 10.1021/acssuschemeng.4c06660

Anokye, K., Mohammed, A. S., Agyemang, P., Agya, B. A., Amuah, E. E. Y., & Sodoke, S. (2024). A systematic review of the impacts of open burning and open dumping of waste in Ghana: A way forward for sustainable waste management. *Cleaner Waste Systems*, *8*, 100152. Advance online publication. DOI: 10.1016/j.clwas.2024.100152

Aromolaran, A., & Sartaj, M. (2024). Microbial Analysis and Methane Assessment of Trinary Anaerobic Co-Digestion of Organic Fraction of Municipal Solid Waste. *BioEnergy Research*, *17*(2), 1250–1262. DOI: 10.1007/s12155-023-10623-5

Babalola, A. A., & Olawuyi, D. S. (2022). Overcoming Regulatory Failure in the Design and Implementation of Gas Flaring Policies: The Potential and Promise of an Energy Justice Approach. *Sustainability (Basel)*, *14*(11), 6800. Advance online publication. DOI: 10.3390/su14116800

Banaś, M., Pająk, T., Bator, J., Wróbel, W., & Ciuła, J. (2024). Energy Efficiency Analysis of Waste-to-Energy Plants in Poland. *Energies*, *17*(10), 2390. Advance online publication. DOI: 10.3390/en17102390

Bhullar, L. (2023). Meanings of energy justice in the low-carbon transition. In *Handbook of Energy Law in the Low-Carbon Transition*. De Gruyter., DOI: 10.1515/9783110752403-011

Biakhmetov, B., Li, Y., Zhao, Q., Ok, Y. S., Dostiyarov, A., Park, Y.-K., Flynn, D., & You, S. (2024). Comparing carbon-saving potential of the pyrolysis of non-recycled municipal plastic waste: Influences of system scales and end products. *Journal of Cleaner Production*, *469*, 143140. Advance online publication. DOI: 10.1016/j.jclepro.2024.143140

Cho, Y., Withana, P. A., Rhee, J. H., Lim, S. T., Lim, J. Y., Park, S. W., & Ok, Y. S. (2024). Achieving the sustainable waste management of medical plastic packaging using a life cycle assessment approach. *Heliyon*, *10*(19), e38185. Advance online publication. DOI: 10.1016/j.heliyon.2024.e38185 PMID: 39398058

Coskuner, G., Jassim, M. S., Jaffer, S., & Ali, S. (2021). Greenhouse Gas Mitigation and Energy Recovery Potential for Optimum Waste Management Scenarios in Bahrain. 2021 3rd International Sustainability and Resilience Conference: Climate Change, 181 – 185. DOI: 10.1109/ IEEECONF53624.2021.9668111

Destrée, P. (2024). Gaseous politics: Contradictions and moral frontiers of the energy transition in Ghana. *Critique of Anthropology*, *44*(3), 235–255. DOI: 10.1177/0308275X241269579

Dong, Y., Zhang, Y., & Liu, S. (2024). The impacts and instruments of energy transition regulations on environmental pollution. *Environmental Impact Assessment Review*, *105*, 107448. Advance online publication. DOI: 10.1016/j.eiar.2024.107448

Ezz, H., Ibrahim, M. G., Fujii, M., & Nasr, M. (2024). Sustainable management of petrochemical wastewater using algal-bacterial granules followed by biogas and biochar production: A techno-economic perspective. *Journal of Water Process Engineering*, *68*, 106391. Advance online publication. DOI: 10.1016/j.jwpe.2024.106391

Fang, M., Njangang, H., Padhan, H., Simo, C., & Yan, C. (2023). Social media and energy justice: A global evidence. *Energy Economics*, *125*, 106886. Advance online publication. DOI: 10.1016/j.eneco.2023.106886

Ghamari, M., Sun, D., Dai, Y., See, C. H., Yu, H., Edirisinghe, M., & Sundaram, S. (2024). Valorization of diverse waste-derived nanocellulose for multifaceted applications: A review. *International Journal of Biological Macromolecules*, *280*, 136130. Advance online publication. DOI: 10.1016/j.ijbiomac.2024.136130 PMID: 39443179

Hai, T., Alenizi, F. A., Alshahri, A. H., Chauhan, B. S., Metwally, A. S. M., & Almujibah, H. R. (2023). Energy and environmental analyses of a sustainable multi-generation municipal solid waste-to-energy integrated system for hydrogen production. *Process Safety and Environmental Protection*, *177*, 307–321. DOI: 10.1016/j.psep.2023.06.030

Haldar, S., Peddibhotla, A., & Bazaz, A. (2023). Analysing intersections of justice with energy transitions in India - A systematic literature review. *Energy Research & Social Science*, *98*, 103010. Advance online publication. DOI: 10.1016/j.erss.2023.103010

Heffron, R. J. (2024). Energy justice–the triumvirate of tenets revisited and revised. *Journal of Energy & Natural Resources Law*, *42*(2), 227–233. DOI: 10.1080/02646811.2023.2256593

Heffron, R. J., & Sokołowski, M. M. (2024). Resolving energy policy failure: Introducing energy justice as the solution to achieve a just transition. *Energy Policy*, *187*, 114042. Advance online publication. DOI: 10.1016/j.enpol.2024.114042

Huang, X. (2024). The Promotion of Anaerobic Digestion Technology Upgrades in Waste Stream Treatment Plants for Circular Economy in the Context of "Dual Carbon": Global Status, Development Trend, and Future Challenges. *Water (Basel)*, *16*(24), 3718. Advance online publication. DOI: 10.3390/w16243718

Ibarra-Esparza, F. E., González-López, M. E., Ibarra-Esparza, J., Lara-Topete, G. O., Senés-Guerrero, C., Cansdale, A., Forrester, S., Chong, J. P. J., & Gradilla-Hernández, M. S. (2023). Implementation of anaerobic digestion for valorizing the organic fraction of municipal solid waste in developing countries: Technical insights from a systematic review. *Journal of Environmental Management*, *347*, 118993. Advance online publication. DOI: 10.1016/j.jenvman.2023.118993 PMID: 37751665

Ibrahim, R. A., Inan, H., & Fahim, I. S. (2023). A comparative cradle-to-gate life cycle assessment of three cotton stalk waste sustainable applications. *Scientific Reports, 13*(1), 20781. Advance online publication. DOI: 10.1038/s41598-023-47817-y PMID: 38012270

International Energy Agency (2023). "Energy Access Report."

Jeetah, P., Somaroo, G., Surroop, D., Ragen, A. K., & Amode, N. S. (2024). Adoption of Machine Learning for Sustainable Solid Waste Management. In *Artificial Intelligence, Engineering Systems and Sustainable Development: Driving the UN SDGs*. Emerald Group Publishing Ltd., DOI: 10.1108/978-1-83753-540-820241002

Karak, S. Garima, Berry, E., & Choudhary, A. K. (2024). Heavy Metal Waste Management to Combat Climate Crisis: An Overview of Plant-Based Strategies and Its Current Developments. In Integrated Waste Management: A Sustainable Approach from Waste to Wealth. Springer Singapore. DOI: 10.1007/978-981-97-0823-9_9

Kaswan, A. (2023). Background on environmental justice and racism. In *The Palgrave Handbook of Global Sustainability* (Vol. 2–3). Springer International Publishing., DOI: 10.1007/978-3-031-01949-4_70

Kaur, A., & Sharma, V. (2024). Nexus between climate change and mitigation approaches for sustainable development: A bibliometric review. *Global Knowledge, Memory and Communication*. Advance online publication. DOI: 10.1108/GKMC-11-2023-0418

Kaur, H., Singh, K., Kumar, P., & Kaur, A. (2023). Assessing the environmental sustainability corridor: An empirical study of Renewable energy consumption in BRICS nation. *IOP Conference Series. Earth and Environmental Science, 1110*(1), 012053. Advance online publication. DOI: 10.1088/1755-1315/1110/1/012053

Longfor, N. R., Aduba, J., Istrate, I.-R., & Qian, X. (2023). Mitigating greenhouse gas emissions from municipal solid waste in Sub-Saharan Africa via sustainable waste management: An economic benefit assessment. *Resources, Conservation & Recycling Advances, 20*, 200192. Advance online publication. DOI: 10.1016/j.rcradv.2023.200192

Marami, H., Tsapekos, P., Adiby, A., Birkved, M., Alvarado-Morales, M., Zhang, Z., Angelidaki, I., Khademi, S., Motevali, A., & Khoshnevisan, B. (2024). To what extent do geopolitical conditions affect the environmental sustainability of the circular economy? A comparative LCA study of lactic acid production in Denmark and Iran. *Sustainable Production and Consumption*, *47*, 363–383. DOI: 10.1016/j.spc.2024.04.006

Marshall, S., & Pearse, R. (2025). Reading the room: Developing a practical justice politics of regional energy transition. *Environmental Politics*, *34*(1), 70–91. DOI: 10.1080/09644016.2024.2347164

Min, Y., & Lee, H. W. (2024). Quantifying clean energy justice: Impact of Solarize programs on rooftop solar disparities in the Pacific Northwest. *Sustainable Cities and Society*, *104*, 105287. Advance online publication. DOI: 10.1016/j.scs.2024.105287

Nawar, K. N., Mahbub, T., Tashfiq, R. A., & Rashid, T. U. (2024). Municipal Solid Waste Collection, Transportation, and Segregation. In Environmental Engineering and Waste Management: Recent Trends and Perspectives. Springer Nature. DOI: 10.1007/978-3-031-58441-1_2

Njangang, H., Tadadjeu, S., & Kamguia, B. (2024). From policy to progress: Do female parliamentarians influence energy justice? *Climate Policy*. Advance online publication. DOI: 10.1080/14693062.2024.2424231

Patil, K. D., De, J., Patil, V. K., & Kulkarni, M. M. (2024). Environmental effects and threats of waste: Understanding threats and challenges to ecosystem, health, and sustainability and mitigation strategies. In *From Waste to Wealth*. Springer Nature., DOI: 10.1007/978-981-99-7552-5_3

Paxton, T. K., Donnellan-Fernandez, R., & Hastie, C. (2023). An exploratory study of women and midwives' perceptions of environmental waste management – homebirth as climate action. *Midwifery*, *127*, 103844. Advance online publication. DOI: 10.1016/j.midw.2023.103844 PMID: 37918131

Perkumienė, D., Atalay, A., Safaa, L., & Grigienė, J. (2023). Sustainable Waste Management for Clean and Safe Environments in the Recreation and Tourism Sector: A Case Study of Lithuania, Turkey and Morocco. *Recycling*, *8*(4), 56. Advance online publication. DOI: 10.3390/recycling8040056

Pillay, R., Hansraj, R., & Singaram, V. S. (2024). Decision-making tree to support sustainable waste management in optometric clinical practice and healthcare training programmes. *Clinical & Experimental Optometry*. Advance online publication. DOI: 10.1080/08164622.2024.2433594 PMID: 39603595

Popa, G., Stegaroiu, I., Robescu, V.-O., Barbu, I., & Ilie, V. (2010). Sustainable development strategy - The key of environmental and organizational management. Proceedings of the International Conference on Energy and Environment Technologies and EquipmProceedings of the International Conference on Energy and Environment Technologies and Equipment, EEETE '10, 88–92. https://www.scopus.com/inward/record.uri?eid=2-s2.0-79952635181&partnerID=40&md5=f939070b8ba44ab91e0a71900ccb9075

Ram, M., & Bracci, E. (2024). Waste Management, Waste Indicators and the Relationship with Sustainable Development Goals (SDGs): A Systematic Literature Review. *Sustainability (Basel)*, *16*(19), 8486. Advance online publication. DOI: 10.3390/su16198486

Ramasar, V., Busch, H., Brandstedt, E., & Rudus, K. (2022). When energy justice is contested: A systematic review of a decade of research on Sweden's conflicted energy landscape. *Energy Research & Social Science*, *94*, 102862. Advance online publication. DOI: 10.1016/j.erss.2022.102862

Rossetto, D. (2023). The relative importance of carbon markets to the waste management sector's future contribution to climate change commitments under the Paris Agreement: Insights from Australia. *Carbon Neutrality*, *2*(1), 25. Advance online publication. DOI: 10.1007/s43979-023-00063-7

Saeed, A., Zafar, M. W., Manita, R., & Zahid, N. (2024). The role of audit quality in waste management behavior. *International Review of Economics & Finance*, *89*, 1203–1216. DOI: 10.1016/j.iref.2023.08.019

Scandinavian Waste-to-Energy Studies (2023).

Shah, H. H., Amin, M., Pepe, F., & Tregambi, C. (2024). Sustainable waste management and waste-to-energy in the context of a circular economy through various waste management technologies. *Environmental Science and Pollution Research International, 32*(48), 27503–27522. Advance online publication. DOI: 10.1007/s11356-024-33223-y PMID: 38589589

Singh, P., Sharma, S., & Dhanorkar, M. (2023). Aquatic plant biomass-derived porous carbon: Biomaterials for sustainable waste management and climate change mitigation. *International Journal of Environmental Science and Technology, 20*(11), 12955–12970. DOI: 10.1007/s13762-022-04601-1

Sovacool, B. K., Baum, C. M., & Low, S. (2022). Climate protection or privilege? A whole systems justice milieu of twenty negative emissions and solar geoengineering technologies. *Political Geography, 97*, 102702. Advance online publication. DOI: 10.1016/j.polgeo.2022.102702

Sumiyati, S., Samadikun, B. P., Widiyanti, A., Budihardjo, M. A., Al Qadar, S., & Puspita, A. S. (2024). Life cycle assessment of agricultural waste recycling for sustainable environmental impact. *Global Journal of Environmental Science and Management, 10*(2), 907–938. DOI: 10.22035/gjesm.2024.02.30

Taşkın, A., & Demir, N. (2020). Life cycle environmental and energy impact assessment of sustainable urban municipal solid waste collection and transportation strategies. *Sustainable Cities and Society, 61*, 102339. Advance online publication. DOI: 10.1016/j.scs.2020.102339

Tereshchenko, E., Happonen, A., Porras, J., & Vaithilingam, C. A. (2023). Green Growth, Waste Management, and Environmental Impact Reduction Success Cases From Small and Medium Enterprises Context: A Systematic Mapping Study. *IEEE Access : Practical Innovations, Open Solutions, 11*, 56900–56920. DOI: 10.1109/ACCESS.2023.3271972

Tisi, Y. S. A. B., Matos, F. A., & Carneiro, M. L. N. M. (2023). Development of waste-to-energy through integrated sustainable waste management: The case of ABREN WtERT Brazil towards changing status quo in Brazil. *Waste Disposal & Sustainable Energy*, *5*(3), 295–308. DOI: 10.1007/s42768-022-00127-x PMID: 36687498

Trakic, A., Karim, R., & Ahmed, P. K. (2024). Energy Justice: What is It and Why It Matters? *Sustainable Development Goals Series*, (Part F3606), 3–10. DOI: 10.1007/978-981-97-6059-6_1

United Nations (2022). "Global Waste Management Outlook."

Varjani, S., Haq, I., Pandey, A., Gupta, V. K., & Bui, X.-T. (2024). Waste Management in Climate Change and Sustainability Perspectives: Organic and Medical Waste. In *Waste Management in Climate Change and Sustainability Perspectives: Organic and Medical Waste*. CRC Press., DOI: 10.1201/9781003386902

Wilson, D. C. (2023). Learning from the past to plan for the future: An historical review of the evolution of waste and resource management 1970–2020 and reflections on priorities 2020–2030 – The perspective of an involved witness. *Waste Management & Research*, *41*(12), 1754–1813. DOI: 10.1177/0734242X231178025 PMID: 37732707

Yap, C. C., Wu, S. L., Soon, P. V., Berezina, E., Aroua, M. K., & Gew, L. T. (2024). Culinary waste management for a healthier planet: A qualitative study. *Cogent Social Sciences*, *10*(1), 2388178. Advance online publication. DOI: 10.1080/23311886.2024.2388178

Yuan, X., Suvarna, M., Lim, J. Y., Pérez-Ramírez, J., Wang, X., & Ok, Y. S. (2024). Active Learning-Based Guided Synthesis of Engineered Biochar for CO2 Capture. *Environmental Science & Technology*, *58*(15), 6628–6636. DOI: 10.1021/acs.est.3c10922 PMID: 38497595

Zhong, F., Tian, J., Zhao, C., Zha, S., Chen, X., & Zhang, Y. (2024). Assessing energy justice in climate change policies: An empirical examination of China's energy transition. *Climate Policy*, *24*(3), 362–377. DOI: 10.1080/14693062.2023.2261894

Chapter 6

Problems and Trends in the Implementation of Digital Transformation of Business Models and ESG Strategies in Corporate Management

Viktoriya Pavlova Kalaydzhieva
https://orcid.org/0000-0002-1825-2228
South-West University "Neofit Rilski", Bulgaria

ABSTRACT

Digital transformation and ESG strategies have a key role in modern corporate management by creating opportunities for sustainable growth and competitiveness. This study examines the functional interdependencies within the triad of digitalization, sustainable development, and modern management. The main research thesis is that digitalization, sustainable development and successful corporate management represent interrelated systemic processes that mutually influence one another. The analysis addresses selected micro-level aspects and explores the relationships between the adoption of digital technologies, the implementation of sustainable development goals, and effective corporate governance. An empirical

DOI: 10.4018/979-8-3373-9685-9.ch006

study based on a survey of 617 respondents from the South-West region of Bulgaria assesses the corporate environment, the readiness for business model transformation, and the current state of key indicators related to sustainable development, corporate governance, and environmental and social practices.

INTRODUCTION

The implementation and use of digital technologies is synonymous with modern innovation processes that guarantee companies potential, competitiveness and strong market positions. Cloud technologies, the Internet of things, big data computing, artificial intelligence are the tools through which companies can change established practices, reform and create new strategies related not only to production processes, but also to communication with business partners and customers. In National Programme "Digital Bulgaria 2025" (2019) and the EU Strategy for the Digital Single Market (European Commission, 2015) the main directions related to the digitalization and digital transformation of business and the social environment are formulated. Digital technologies are transforming not only business but also all spheres of social relations – both personal and professional.

Sustainable development is a contemporary and important global project with long-term implications for the future of the world. It is a continuous systemic process that requires innovative and complex changes. Its complexity stems from the fact that it affects three interconnected subsystems, and the problem must be addressed simultaneously because of their interdependence. The Report of the World Commission on Environment and Development (United Nations, 1987), also known as Brundtland Report, did not outline an algorithm or plan of action. „Increasing attention is also being paid to corporate sustainability, with research focusing on how businesses establish and sustain connections with the pertinent parties like suppliers, agencies, shareholders and so on, as well as how management and staff embrace and implement sustainability strategies" (Filipova et al., 2024). Many questions arise: Should push or pull factors dominate the implementation of changes related to sustainable development? Are the factors causing resistance to transformation fully understood? Are all countries ready, and to what extent is this project adequately resourced? Are the changes occurring at an appropriate pace, and how efficient are they? The answers to these questions require the application of various research methods and approaches - historical, systematic, comparative, interdisciplinary, inductive, deductive, analytical, synthetic, and statistical. The issue concerns the company, national, and global levels in order to achieve real action and timely results.

THEORETICAL ASPECTS AND RESEARCH BACKGROUND

The terms "digitization", "digitalization", "digital transformation" and "green digitalization" have a common etymology, but different semantics. Frenzel et al. (2021) provide a systematic review of the essential characteristics and differences of the concepts digitization and digitalization. According to them, despite the existing solid foundation, until now there is a lack of a generalized concept on this issue. According to Ross (2017) there is a lack of common understanding related to "digitization" and "digitalization", both in the literature and in academic circles. Authors such as Legner et al. (2017) make a clear distinction and differentiation between both concepts. According to other authors (Lyytinen et al., 2016), such a significant difference does not exist. In recent years, the opinion that digitization affects product architectures, with the final stage representing a platform-ecosystem, has been increasingly asserted (Sandberg et al., 2020). According to Brennen and Kreiss (2014, 2016), "digitization" and "digitalization" are two close conceptual terms that are often used interchangeably by several authors in the scientific literature. The term "digitalization" is associated with the change and restructuring of many spheres of social life using digital communication and media infrastructures. Through "digitalization", international business is carried out in a new way, which provides opportunities for optimization and efficiency of contacts with partners in the internationalization of business relations. According to Yuleva-Chuchulayna, in the context of globalization and digitalization, enterprises are faced with the constant maintenance of organizational advantages over other companies in theconditions of strong competition and an open market (Yuleva-Chuchulayna, 2024). Modern digital technologies such as cloud technology, big data, artificial intelligence, Internet of things, blockchain technologies, virtual engineering, neural networks, etc. and the opportunities they provide are associated with the term "digital transformation". The concept of "green digitalization" is associated with sustainable development and its environmental aspects. In this study, the term "digitalization" is used as a synonym for the implementation of modern digital technologies, which are considered as a tool for achieving digital transformation. Depending on the degree and scope of their influence on the internal and external environment of the company, it can be determined whether the respective business has achieved digital transformation. Digital technologies nowadays are a reality not only in the business environment, they are the modern communication channels in society as a whole. Digital transformation is a change not only in the field of economics, but also a transformation of communication. According to Stefanova et al. (2023) the current dynamics and crises in the modern environment seriously affect the organizational-functional structures and management. Innovation processes are evolving and the focus is now on new innovative practices that build on proven methods. Effective communication is a fundamental tools for addressing the challenges posed by environmental pressures.

Digital transformation as an organizational change is the subject of research by many authors, institutes and consulting companies such as Institute for Management Development (IMD) and SAP (2016). The essence of the category "digital transformation" is defined by Wade (2015) and IMD as a plan that includes 7 main components - how you make money; how you are organized; the people who work for you; how you do things; how you collect and manage information; your products and services; how you engage with customers, suppliers, etc. The additionally formulated guiding questions set an algorithm for the stages in implementing digital transformation at the company level. Digital transformation requires changes in the company strategy, company values and culture.

When using modern digital technologies, all the negatives and risks must be considered. The initiatives and regulations within the European Union related to the impact of artificial intelligence should be noted. The preparation of harmonized rules "Artificial Intelligence Act" on the use of artificial intelligence is a step in this direction. Such actions are particularly needed in sectors with social or environmental impact, including climate change, environment and health, the public sector, finance, mobility, home affairs and agriculture. It is in the interest of the Union that new technologies, which are developed and operational, align with the Union's values, fundamental rights and principles (European Parliament, 2021). A priority in the European Commission's policy is to create and implement legislation that guarantees a common approach for European countries in the use of artificial intelligence, with an emphasis on ethical norms and relationships in society. The adoption of a National Strategy and Roadmap for Digital Transformation, corresponding to the European goals, is extremely important for the business policies of every country and company.

Until recently, artificial intelligence was associated more with its hardware essence and forms such as assembly lines, robots and other cyber-physical devices. Today, its software "manifestations" are an inevitable and irresistible part of the daily life of businesses and people. Even organizations with no affinity for using digital technologies are forced by "push" factors to change because consumers are already "digitalized", and the same applies to investors, suppliers and counterparties. Undoubtedly, monitoring, analysis, forecasting, control and implementation of all activities related to the eclogical aspect of sustainable development are unrealizable without digital technologies, which are now rightfully called "green digitalization".

Existing negative attitudes towards sustainable development often lead to rejection of transformational actions, especially at the company level. Small and medium-sized enterprises (SMEs) do not have the necessary technological infrastructure. Even many of the larger companies lack a vision and a strategy for implementing digital transformation and applying ESG (environmental, social and governance) practices. The push factors imposed by the external environment, science and the market prevail, which act as coercive incentives. Aggressive market competition is often cited as a reason for compromises regarding the application of ESG practices. The financial and other resources required for their implementation are not small and quickly repaid investments, which is why many companies stick to established practices and imposed stereotypes. Many companies lack sufficiently qualified personnel, good company systems for selection and training of employees, as well as the opportunity for continuing professional training, for evaluating results and career development and growth.

The insufficient definition of ESG practices leads to problems with motivation and difficulties in measuring and evaluating results from an environmental, social and economic aspect. A new control and accountability model and infrastructure is needed that sets clear rules, as is the case with financial reporting. A new culture, a new way of thinking, a new worldview must be created to overcome the classic conflict between the pursuit of profit and sustainability, in its ecological and social aspects. It is imperative to understand and accept that digital transformation and ESG activities are not an expense, but an investment that will yield a positive results and returns in the future. Sustainability is a complex functional dependency and its assessment must change to take this fact into account. The current assessment, selection of indicators and metrics are ineffective. At the micro and macro level, the main issue is related to ensuring competitiveness and achieving sustainable economic growth.

Sustainable competitive advantage can be applied to a company, country, or individual. It is mainly of three types-comparative advantage, cost advantage, and differentiation advantage. The strategies with which an organization can earn are: information advantage; cost leadership strategy; adaptability; differentiation strategy; operational effectiveness strategy; technology-based competitive advantage; and innovative strategy (Serafimova & Vasilev, 2024, p.220).

Wide connectivity greatly improves the effectiveness and outcomes of social innovations. A number of studies have found that the social pillar of sustainable development is perceived and implemented by companies, unlike the environmental pillar, which is associated with much more investment and risks. Through digital technologies, information related to poverty and violence can be collected and processed. Digital technologies are successfully applied today in areas such as agriculture, water basin management and forestry. The term "green digitalization" has quite rightly appeared as an analogue of the concept of green economy. Artificial intelligence and blockchain technologies are involved in the creation of smart energy systems, cities and homes, digital transformation provides opportunities to change the previous forms of workplace organization, time, performance evaluation and remuneration. Education and continuing professional training are facilitated and modernized through new digital resources. Opportunities for awareness and disclosure through digital communication channels have a key role in many social and environmental problems and crises. Digital re-use markets are already functioning. Attitudes and preferences change through social networks, which enables companies implementing eco-standards to reach new markets and customers. The capabilities of digital technologies are key factor for the transformation and modernization of the public sector – security of personal and company data, cyber security, improving the accountability of institutions, modernization and facilitation of administrative services, providing access to platforms for electronic consultation on a number of health and public issues. Virtual engineering, with its possibilities to create digital models, simulations and prototypes, not only seriously reduces the consumption of energy, materials, human resources, but also at the same time reduces harmful effects on the environment. Digital technologies enable the analysis and extension of the life cycle of products and services. This leads to reuse, minimizes resource costs and supports the circular economy.

EUROPE IN THE DIGITAL AGE AND EUROPEAN GREEN DEAL

Regulated by the European Commission, the Digital Economy and Society Index (DESI) is related to the progress member states in the field of digital technologies. It annually identifies and provides opportunities for analysis and assessment of the current state, trends and dynamics of processes related to digitalization in the EU countries. The following two tables provide data related to the four main indicators of Digital Economy and Society Index (European Commission, 2024).

Table 1. DESI 2024 Indicators - Digital Skills and Digital Infrastructure

Country	*DESI 2024 (data fom 2023)*			
	DESI indicators			
	Digital skills		**Digital infrastructure**	
	Above basic digital skills %	**Internet use %**	**Overall 5G coverage %**	**Fibre to the Premises (FTTP) coverage %**
European Union - 27 countries (from 2020)	**27.32**	**90.27**	**89.30**	**63.99**
Austria	32.04	91.99	96.03	40.97
Belgium	28.26	93.57	40.35	25.00
Bulgaria	7.73	79.83	70.88	88.61
Cyprus	24.96	91.19	100.00	77.07
Czechia	35.48	89.91	94.60	36.05
Germany	19.79	90.98	98.14	29.80
Denmark	39.37	97.47	100.00	84.04
Estonia	34.84	92.29	87.47	76.94
Greece	20.02	84.23	98.07	38.41
Spain	38.65	94.48	92.28	95.21
Finland	53.63	95.45	98.35	61.15
France	30.62	92.17	93.21	81.40
Croatia	25.00	82.46	83.43	62.09
Hungary	28.13	90.63	83.70	76.16
Ireland	43.82	93.17	85.32	78.47
Italy	22.21	85.45	99.49	59.29
Lithuania	25.90	87.59	98.87	78.06
Luxembourg	27.86	96.87	99.60	78.90
Latvia	16.55	91.50	53.11	61.88
Malta	36.98	91.51	100.00	69.60
Netherlands	54.53	98.92	100.00	77.66
Poland	20.05	85.32	71.92	75.43
Portugal	29.93	84.18	98.09	92.32
Romania	8.97	88.05	32.75	94.99
Sweden	36.51	96.87	90.28	83.92

continued on following page

Table 1. Continued

Slovenia	18.88	89.41	82.13	78.51
Slovakia	21.70	86.06	79.03	64.19

Source: Adapted from European Commission. (2024)

For each of the indicators listed in Table 1, the number of countries with lower than the European average values is respectively: 12 (44.44%); 11 (40.74%); 11 (40.74%) and 9 (33.33%). The Netherlands, Denmark, Spain and Finland have leading positions in most of these four indicators. Lagging behind in many respects, Romania and Bulgaria have good results in Fibre to the Premises (FTTP) coverage.

Table 2. DESI 2024 Indicators - Digital Transformation of Business and Digitalization of Public Services

	DESI 2024 (data fom 2023)						
	DESI indicators						
	Digital transformation of businesses					Digitalisation of public services	
Country	**Data analytics %**	**Cloud %**		**Artificial intelligence %**		**Digital public services for citizens %**	**Digital public services for businesses %**
European Union - 27 countries (from 2020)	**33.20**	**38.90**		**8.00**		**79.44**	**85.42**
Austria	23.90	35.60		10.80		80.72	82.86
Belgium	44.50	47.70		13.80		82.33	91.59
Bulgaria	21.90	14.20		3.60		67.47	**91.88**
Cyprus	33.50	45.50		4.70		73.96	86.07
Czechia	19.50	35.20		5.90		76.33	83.75
Germany	37.10	38.50		11.60		75.83	78.58
Denmark	49.50	66.20		15.20		84.24	88.69
Estonia	25.60	52.60		5.20		95.83	98.75
Greece	25.00	18.10		4.00		75.92	86.20
Spain	38.00	27.20		9.20		84.18	91.00
Finland	40.60	73.00		15.10		90.61	100.00
France	33.90	22.90	b	5.90	b	72.09	79.31
Croatia	51.70	40.70		7.90		67.17	66.18
Hungary	53.20	37.10		3.70		73.36	74.86
Ireland	37.10	53.10		8.00		81.21	100.00
Italy	26.60	55.10		5.00		68.28	76.27

continued on following page

Table 2. Continued

Lithuania	40.50	33.60		4.90		86.70	95.94
Luxembourg	32.40	32.60		14.40		94.78	96.67
Latvia	36.90	29.00		4.50		88.22	87.22
Malta	35.60	58.20		13.20		100.00	100.00
Netherlands	48.60	57.40		13.40		85.87	86.67
Poland	19.30	46.50		3.70		63.73	72.88
Portugal	38.60	32.30		7.90		81.54	81.94
Romania	21.90	15.50		1.50		52.18	50.03
Sweden	35.00	66.00	b	10.40	b	93.28	95.97
Slovenia	19.10	36.00		11.40		77.00	83.96
Slovakia	30.20	30.20		7.00		72.06	79.18

Note. (b) break in time series.
Source: Adapted from European Commission. (2024)

The use of Cloud technologies in nearly 56% of the countries is below the European average. In the case of Artificial intelligence, these are approximately 52% of the countries. For the indicator "Digital public services for citizens", there are 48.1% of such members of the Union. Regarding the indicators from the fifth and first columns of Table 2, this concerns 44.44% and 40.74% of countries below the average Europen values of these indicators, respectively. If we compare the data in Eurostat for 2024 with the corresponding results from 2023, it will be seen that in most cases the trends are positive, but the dynamics are not good enough. The countries with the best values for each of the four indicators for the previous year retain their leading positions in 2024 as well. The same can be said for the countries lagging behind the average European level. The conclusion has to be drawn that Europe's digital structure is good, but a higher speed is needed in implementation of digitization, digitalization and digital transformation. A number of countries seriously lag behind the average European levels in many of the sub-indicators. The idea of a greener, more digitalized and sustainable Europe is enshrined in Recovery plan for Europe (European Commission, 2020). "The fight against to the climate change requires the implementation of multifaceted policies. Climate change is not just about environmental sustainability; It also causes negative effects on social welfare and economic continuity" (Yıldırım & Vasilev, 2024, p.2). Legislation related to improving human's well-being and health and economic prosperity in a sustainable way for our planet aims to make the European Green Deal (European Commission, 2019) a reality. 600 billion euro of dedicated investment funds are foreseen through the NextGenerationEU Recovery plan, zero emissions from new cars by 2035 and 40% is the new target set for renewable energy by 2030 (European Commission, 2021a). The serious impact on the environment is a continuous trend, which is influenced by industrial growth, increased energy demand and population growth. These factors lead to significant ecological problems and challenges that require urgent and adequate action.

The EU's commitment to achieve climate neutrality by 2050 is regulated by the European Climate Law (European Commission, 2021b). Europe is becoming a world leader in the fight against climate change, aiming to limit global warming to 1.5°C. The Green Deal Industrial Plan of February 2023 envisages the creation of 4.5 million green jobs to increase the competitiveness of net-zero industry and to make the transition to climate neutrality. The goal is by 2030 the deployment of clean strategic technologies for net zero emissions to exceed 40%. An increase in renewable energy production capacity is foreseen, and the improvement in energy efficiency should reach 11.7% by 2030 (European Commission, 2021a, 2023).

The circular economy is a key element in sustainable development. The following table shows data on waste generation per capita (according to the latest published data in Eurostat).

Table 3. Waste Generation per Capita

Country	**2018**		**2020**		**2022**	
Kilograms per capita						
European Union - 27 countries (from 2020)	**5 235**		**4 817**		**4 991**	
Belgium	5 967		5 899		5 363	
Bulgaria	18 470		16 785		14 603	
Czechia	3 560	b	3 598		3 672	
Denmark	3 702		3 453		3 333	
Germany	4 891		4 824		4 604	
Estonia	17 539		12 163		16 752	

continued on following page

Table 3. Continued

Country	2018		2020		2022	
Ireland	2 874		3 248		2 971	
Greece	4 215		2 651		2 858	p
Spain	2 945		2 230		2 480	
France	5 112		4 577		5 076	
Croatia	1 355		1 483		1 838	
Italy	2 855		2 942		3 212	
Cyprus	2 646		2 491		3 294	
Latvia	920		1 501		1 330	
Lithuania	2 527		2 396		2 003	
Luxembourg	14 828		14 618		15 169	
Hungary	1 879		1 759		2 838	
Malta	5 336		6 938		5 004	
Netherlands	8 429		7 175		6 921	
Austria	7 428		7 728		8 079	
Poland	4 621		4 492		4 739	
Portugal	1 546	b	1 612		1 878	
Romania	10 425		7 344		8 410	
Slovenia	3 964		3 576		5 397	
Slovakia	2 277		2 338		2 462	
Finland	23 253		20 993		19 950	
Sweden	13 628		14 664		15 627	

Note. Special value: (:) not available; (b) break in time series; (d) definition differs (see metadata); (p) provisional.

Source: Eurostat. (2024f)

If we compare the data from Table 3 with the data from Eurostat 2024 for Recycling rate of all waste excluding major mineral waste, Recycling rate of packaging waste by type of packaging and Recycling rate of waste of electrical and electronic equipment (WEEE) separately collected, countries can be categorized into the following four groups, which do not exhaust the complete list of countries.

Group A. Countries that generate little waste and at the same time have good indicators (above the European average) related to recycling – Belgium, Germany, Denmark, Czechia, Slovakia, Italy.

Group B. Countries that generate little waste, but their indicators related to recycling are worse than the European average – Ireland, Greece, Spain, Cyprus, Hungary.

Group C. Countries that generate waste above the average European level but have good indicators related to recycling – Luxembourg, Finland, Estonia.

Group D. Countries that generate a lot of waste and at the same time have poor practices and results related to recycling – Liechtenstein and Bulgaria, with the proviso that in terms of Recycling rate of waste of electrical and electronic equipment separately collected, Bulgaria after 2020 has values above the average European level by several percentage points.

Regarding the indicators Generation of municipal waste per capita and Recycling rate of municipal waste, by analogy with the indicator Waste generation per capita, four groups of countries can also be generated. Countries such as Slovakia, Italy, Germany and the Netherlands have the best values of these indicators – better or close to the European average. In countries such as Poland, Romania, Estonia and Bulgaria, there is a discrepancy between the values of these two indicators – low levels of municipal waste generation, but also a low degree of material recycling.

In Eurostat (2024b) can be found data related to the indicator Consumption footprint (Table 4). The indicator Consumption footprint estimates the environmental impacts of EU and Member States consumption by combining data on consumption intensity and environmental impacts of representative products.

Table 4. Consumption Footprint

Country	2020	2021	2022
	Index, 2010=100		
European Union - 27 countries (from 2020)	**103**	**106**	**109**
Belgium	107	110	108
Bulgaria	109	114	117
Czechia	103	111	115
Denmark	112	112	117

continued on following page

Table 4. Continued

Country	2020	2021	2022
	Index, 2010=100		
European Union - 27 countries (from 2020)	**103**	**106**	**109**
Germany	105	102	104
Estonia	104	104	107
Ireland	94	86	90
Greece	88	92	97
Spain	97	107	116
France	100	104	107
Croatia	109	115	118
Italy	95	96	103
Cyprus	100	105	113
Latvia	99	98	105
Lithuania	103	108	108
Luxembourg	122	130	134
Hungary	106	108	110
Malta	133	137	144
Netherlands	96	98	105
Austria	103	105	107
Poland	117	122	123
Portugal	107	110	118
Romania	107	111	114
Slovenia	102	99	103
Slovakia	111	112	119
Finland	96	98	101
Sweden	87	91	100

Source: Eurostat. (2024b)

The data in the above table confirm the findings in the Sustainable development report (2023) about the lack of dynamics of the processes, and in many cases, negative trends related to this aspect of sustainable development.

Table 5 shows the countries that achieved results better than the average European levels for the period 2021 - 2023 (according to the latest published data in Eurostat), in terms of circular material use.

Table 5. Circular Material Use Rate

Country	2021	2022	2023	
	% (Percentage)			
European Union - 27 countries (from 2020)	**11.1**	**11.5**	**11.8**	s
Belgium	21.40	18.3	19.7	e
Czechia	11.00	11.3	12.8	s
Germany	12.20	12.5	13.9	s
Estonia	20.00	21.4	18.1	s
France	15.90	17.5	17.6	s
Italy	19.70	20.6	20.8	s
Malta	19.20	21.5	19.8	s
Netherlands	28.50	27.2	30.6	s
Austria	11.30	12.4	14.3	s
Sweden	9.50	12.1	9.9	s

Note. (s) Eurostat estimate; (e) estimated.
Source: Adapted from Eurostat. (2024a)

Significantly below the European average for 2023 are the results of Romania (1.3%), Ireland (2.3%), Finland (2.4%), Portugal (2.8%), Lithuania (3.9%), Bulgaria (4.9%) and Greece (5.2%). The indicator measures the share of material recycled and fed back into the economy - thus saving extraction of primary raw materials - in overall material use. The circular material use, also known as circularity rate, is defined as the ratio of the circular use of materials to the overall material use. A higher circularity rate value indicates means that more secondary materials substitute for primary raw materials thus reducing the environmental impacts of extracting primary material.

Useful and objective information about the real commitment of the EU countries to the policies and processes related to the implementation of the circular economy is contained in the following table.

Table 6. Persons Employed in the Circular Economy Sectors and Population on 1 January by Age and Sex

Country	2021	2022	2023		Population on 1 January by age and sex for the year 2023	
	Number of persons employed	Number of persons employed	Number of persons employed	Employed in the circular economy %	Number	
European Union - 27 countries (from 2020)	**4 704 650**	**4 248 173**	**4 368 611**	**0.97**	**448 803 078**	**ep**
Belgium	70 371	61 965	61 760	0.53	11 742 796	
Bulgaria	102 385	94 319	95 233	1.48	6 447 710	
Czechia	153 934	137 340	137 827	1.27	10 827 529	
Denmark	47 132	42 049	41 556	0.70	5 932 654	
Germany	783 930	731 414	771 814	0.91	84 358 845	
Estonia	27 113	48 923	54 510	3.99	1 365 884	
Ireland	44 945	27 874	36 056	0.68	5 271 395	
Greece	97 477	79 328	76 522	0.73	10 413 982	
Spain	460 871	402 408	428 345	0.89	48 085 361	
France	597 322	533 892	537 036	0.79	68 172 977	p
Croatia	75 346	66 553	66 309	1.72	3 850 894	
Italy	578 336	515 093	507 749	0.86	58 997 201	
Cyprus	13 514	13 464	13 849	1.50	920 701	
Latvia	33 425	31 373	32 971	1.75	1 883 008	
Lithuania	63 976	58 707	61 452	2.15	2 857 279	
Luxembourg	8 265	19 435	21 482	3.25	660 809	
Hungary	130 399	108 280	109 363	1.14	9 599 744	
Malta	9 700	7 883	8 145	1.50	542 051	
Netherlands	119 839	108 104	110 564	0.62	17 811 291	
Austria	71 734	65 022	65 849	0.72	9 104 772	
Poland	435 868	401 558	411 141	1.12	36 753 736	ep
Portugal	139 375	119 809	117 372	1.12	10 516 621	
Romania	239 087	200 489	198 459	1.04	19 054 548	e
Slovenia	30 686	27 241	29 463	1.39	2 116 972	
Slovakia	67 492	63 216	64 966	1.20	5 428 792	
Finland	45 575	43 543	41 773	0.75	5 563 970	
Sweden	87 963	74 235	79 323	0.75	10 521 556	
United Kingdom	549 618	556 802	563 692	:	:	

Note. (:) not available; (e) estimated; (ep) estimated, provisional; (p) provisional.
Source: Adapted from Eurostat. (2024d). Eurostat (2024e)

The percentage dimensions of this indicator in Table 6 were calculated by the author, with the population data for each country taken from Eurostat 2023. This was done because the information published in Eurostat about the number of persons employed in the sectors of the circular economy is not related to the population of the respective country. For countries such as Luxembourg, Cyprus and Malta, the quantitative data are minimal, but in percentage units per capita, these countries are among the leaders. The same can be said about the absolute leader Estonia.

Table 7 shows the countries with indicators better than the average European level in the period 2021-2023 related to greenhouse emissions (according to the latest available data in Eurostat).

Table 7. Greenhouse Gases Emissions from Production Activities

Country	2021		2022		2023	
	Kilograms per capita					
European Union - 27 countries (from 2020)	**6541.05184**	s	**6464.89965**	s	**5965.43833**	s
Spain	4854.93283		5001.0581		4590.56515	e
France	4856.51969		4734.20722		4424.18974	
Croatia	4711.22967		4735.80367		4515.01053	s
Italy	5271.48859		5347.19028		4990.41437	e
Latvia	5212.54314		5187.02573		5236.37822	s
Hungary	5241.0926		5102.88558		4806.93622	s
Portugal	4717.48654		4871.76243		4507.00789	s
Romania	5003.02501		4748.49016		4660.57238	s
Slovenia	6069.06711		5535.97732		5285.51219	s
Slovakia	6290.96487		5556.46952		5583.14208	s
Sweden	4074.10223		3970.88565		3833.55266	

Note. (s) Eurostat estimate; (e) estimated.
Source: Adapted from Eurostat. (2024c)

For the remaining countries from EU - 27, the values of these indicators for 2023 range to 16 930. The worst values are those of Iceland (16 930), Denmark (12 481), Luxembourg (11 332) and Ireland (11 176).

The Special Report of the European Court of Auditors (2023) on the circular economy found that progress towards a circular economy is slow; the contribution of projects to support access to modern production technologies is limited. The Sustainable Development Report (2023) provides an analysis and findings related to the current state, problems and challenges related to the Global Goals. The main finding and recommendation is to improve the dynamics of the processes related to the implementation of the main goals. The Sustainable Development Report (2023) proposes a plan to reform the financial framework and national strategies for the global goals, as well as long-term investment intentions and concrete actions. European integration and policies support sustainable development that meets the needs of the present generation without compromising the ability of future generations to meet their own needs (Yuleva-Chuchulayna & Filipova, 2023).

The research of Schaltegger et al. (2002) provides specifics for integrating sustainable development at the company level by grouping the challenges into four groups, adding corporate sustainable development as part of an integrity of ecological-economic and socio-economic givens.

DIGITALIZATION AND ESG PRACTICES IN THE SOUTH-WESTERN BULGARIA. SURVEY "DIGITAL TECHNOLOGIES, SUSTAINABLE DEVELOPMENT AND MODERN MANAGEMENT"

In the context of serious changes in European policies related to global green digitalization and Bulgaria's lagging behind in many indicators in this regard, it was of interest to the author to study these processes and problems at the regional company level. The survey was conducted by the author in the period January-March 2024, the sample included 617 respondents, including 253 owners, managers and team leaders of companies and organizations in South-Western Bulgaria. The goal is to clarify what is the environment, understandings and current state of some basic parameters related to digital transformation, sustainable development and modern management in the southwestern region. In relation to the research question and the set goal, some preliminary hypotheses and assumptions have been formulated that influenced the preparation of the questionnaire. The design of the survey card includes four groups of questions related to: data about the respondent and his organization; level of digitalization of the company; implementation of sustainable development and modern management. The sample as a whole has heterogeneous nature, aiming to seek a wide range of respondents, perspectives and opinions, which positively affects the credibility and representativeness of the information obtained.

The individual groups of questions make it possible to search for and analyze correlations in the triad "digitalization - sustainable development - modern management". The essential characteristics of the studied objects, explicit and indirect functional dependencies between them were analyzed. The first 10 questions in the survey determine the profile of the respondents and their organizations. Of the total number of 617 people, 60.8% are women and 39.2% are men. The distribution by age groups resembles a normal distribution and grouping around the average age (36-55 years). People holding management positions (team leader, manager/director and owner) are 253 people, which is 41% of the sample, which was included in the preliminary design, in order to provide representative opinions and information on the research questions. The remaining 59% of the respondents provide a good control quota of the leadership opinion and ensure a wide range of ideas and opinions. The largest share is the companies established more than 10 years ago – 78.9%, while the number of the rest is evenly distributed in the other time periods (1-3, 4-7 and 8-10 years). Only 5.2% of the respondents do not know when their company was established. 27.2% have minimal experience, and 46.0% have the longest experience. In the distribution by "activity carried out by the organization", the largest share is the Financial, IT and Educational sectors, as well as Commerce and Services, which is expected and logical considering that the South-West region, including the capital Sofia, gives nearly 49% of the Gross Domestic Product (GDP) of the country. In terms of the number of employees in the organization, the sample is relatively evenly distributed - all groups of enterprises are covered. Companies without foreign participation prevail - 62.2%, but the share of those with foreign participation is also not small - 37.8%. The management system is defined as "strictly centralized" by 73.7%.

The next group of questions (11-29) are related to digitalization, education, qualification and continuing training. 63.7% responded positively to the question about the availability of sufficient investments for digitalization of activities. Only 7.9% answered that the investments are extremely insufficient, and another 20.7% defined them as insufficient. An IT department operates in 61.9% of companies, and these departments are most often small (up to 3 people) or large (more than 10 people). The correspondents evaluate the level of digital competences and skills of their colleagues as *"low", "average", "good" and "very good"* with indicators of 8.3%, 23.8%, 41.0% and 26.9% respectively. The fact that 72.3% (446 people) need courses and training related to the use of new digital technologies is disturbing. Establishing relationships with partners and customers through digital technologies was found in 73.1% of respondents. The fact that "face-to-face" contact is the most widespread at 76.0%, along with "e-mail" (80.4%) brings clarity to the previously considered question. Namely, that digital contacts with partners and customers may exist, as 451 respondents claim, but they are mainly limited to the use of e-mail and smart phones.

Some of the questions related to the digital transformation of business and the answers sought were induced by the Institute for Management Development guiding questions related to the category of digital transformation. The degree of digitalization and automation by sector shows that digitalization is assessed as "*very good*" in the sectors of Information and Communication Technologies, Transport, Finance, Banking and Insurance and Healthcare. The lowest scores are in the sectors of Construction, Agriculture and Light and Food Industry. The degree of digitalization is determined as "*good*" and "*very good*" by 64.0% of respondents, with the largest relative share of those from large enterprises - 77.8% (172 respondents from 221 representatives of enterprises (250+ people)). The indicator "degree of automation and digitalization" of companies with foreign participation, assessed as "*very good*" has significantly higher percentage values than the corresponding ones for Bulgarian companies - 42.9% to 16.9%, calculated based on the number of companies of the respective type.

The problems related to sustainable development at the national level were also confirmed at the level of the South-West region. The hypothesis that people associate sustainable development mainly with the economic pillar of sustainable development was confirmed – 66% associate it with competitiveness. Eco-standards and social justice are associated with sustainable development by fewer people. The basis for such a conclusion is that 53.0% of respondents do not know the term "circular economy", 70.5% declare non-use or lack of information about ESG practices. Only 15.4% associate sustainable development with the use of recycled materials. This is probably due to unfamiliarity with the semantic essence of the term sustainable development, circular economy and ESG practices, because a significantly larger percentage of respondents declare the existence of practices related to material recycling (36.6%), waste management (36.6%) and the use of second-hand products (16.9%). The dominant answer to this question is energy saving. What is more disturbing is that 21.4% of respondents declared not implementing any of the above-mentioned practices.

The quality of the organization's human resources was assessed as "*good*" and "*very good*" by 49.3% and 22.5% of the respondents, respectively. Regarding the existence of clearly defined rules and systems for evaluation and career growth, a definite answer *yes* was given by 26.9% of respondents. A negative answer ("*no*" and "*rather not*") was given by 34.8%, i.e. from 215 respondents.

Table 8. Managers and Owners (156 people) for the Term "Sustainable Development"

What do you associate the term "sustainable development" with? (you may indicate more than one answer)	**Total number**	**Total %**
Competitiveness	108	69.2
Serious market share	64	41.0
High turnover and profit	45	28.8
Compliance with environmental standards	35	22.4
Specialized products	22	14.1
Quality and low prices	27	17.3
Customer satisfaction	98	62.8
Social justice	39	25.0
Use of recycled materials	21	13.5
Circular economy	29	18.6
Resource-efficient production	22	14.1
Total:	**510**	**326.9**

Source: Own survey

An opportunity was given to indicate several answers to this question. The values in the second column are a percentage of 156, which is the number of owners and managers/director. Owners and senior management associate sustainable development mostly with the economic pillar – competitiveness, serious market share, high turnover, profit and customer satisfaction. Social justice is associated with sustainable development in 35.6% of large enterprises (250+ people).

The table below provides information on the knowledge of ESG practices by leaders in the companies, as well as on their implementation.

Table 9. Managers and Owners for ESG Strategy or Practices Related to Environmental, Social and Corporate Governance

Does your organization have an ESG strategy or practices related to environmental, social and corporate governance?	Position in the organization			
	Manager/Director	Owner	Total	Total in %
Yes	32	17	49	31.4
No	29	57	86	55.1
I have no information	11	10	21	13.5
Total:	**72**	**84**	**156**	**100**

Source: Own survey

A large part of the surveyed owners and managers/directors answered that there is no ESG strategy, practices or they have no information about it in their organization – 79.8% and 55.6% respectively. This also applies to 69.1% of team leaders. We would like to clarify that the abbreviation ESG was described in the question as related to environmental, social and corporate governance. The total number of those who answered "*no*" and "*I have no information*" is 435 out of a total of 617 respondents. Of the 253 executives surveyed (including team leaders), 110 responded with "*no*" and 64 with "*I have no information*". Therefore, 68.8% gave a negative answer. The information in the above two tables is disturbing and shows the lack of traditions, awareness and attitudes on the discussed topic. This conclusion is also confirmed by the data in the following Table 10 and Figure 1.

Table 10. Implementation of "Green" Practices in the Work of the Organization, according to when the Organization was Established

When your organization was established?	Are "green" practices applied in the organization's work?					
	Yes	**No**	**I have no information**	**Rather yes**	**Rather not**	**Total**
More than 10 years	114	84	54	128	107	487
I have no information	2	9	9	8	4	32
From 1 - 3 years	8	8	6	7	7	36
From 4 - 7 years	7	4	5	10	7	33
From 8 - 10 years	8	5	2	6	8	29
Total:	**139**	**110**	**76**	**159**	**133**	**617**

Source: Own survey

In enterprises established more than 10 years ago, the positive answers are 49.7%, i.e. more than half of the answers are negative or related to "*I have no information*". The indicators are similar or worse in the other groups of enterprises. The exception are the organizations that have been established from 4-7 years, where 51.5% of respondents answered positively.

Figure 1. Knowledge of the Term "Circular Economy", according to whether the organization has or does not have foreign participation

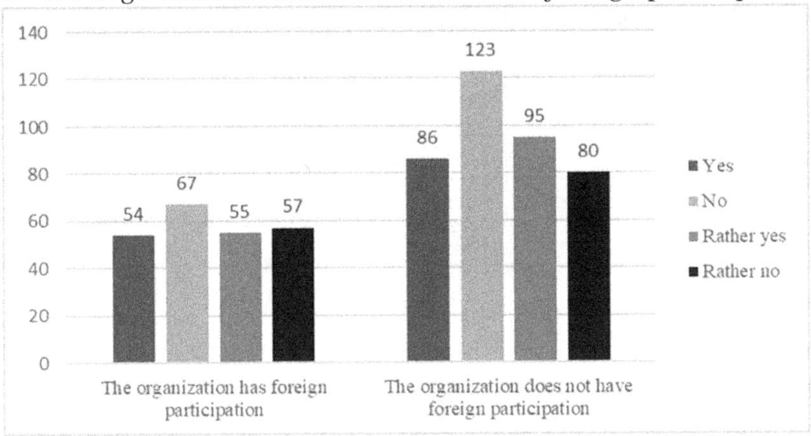

Source: Own survey

Almost two times less representatives of companies with foreign participation answered "*no*" compared to those of Bulgarian companies without foreign participation, regarding the knowledge of the term "circular economy".

The implementation of "green" practices is characteristic mainly of large companies with foreign participation. In Bulgarian companies without foreign participation, 51.6% of the respondents answered "*no*" and "*rather not*", while in companies with foreign participation this value of the indicator is 19.3%. The difference is quite significant, which shows a serious lag in terms of awareness, behavior and attitudes towards one of the three pillars of sustainable development in the country and in particular in the region. From the survey it can be concluded that in most sectors "green practices" are not applied or there is no information about them. The author's preliminary hypothesis that in organizations with a low degree of digitalization, the implementation of ESG practices is also unsatisfactory, was confirmed.

Table 11. Implementation of ESG Strategy or Practices Related to Environmental, Social and Corporate Governance in Organizations with Good and Very Good or Low Degree of Automation and Digitalization

What is the degree of automation and digitalization of your organization's activities?	Does your organization have an ESG strategy or practices related to environmental, social and corporate governance?			
	Yes	No	I have no information	Grand Total
Good and Very good	144	95	156	395
Low	8	34	18	60
Total:	**152**	**129**	**174**	**455**

Source: Own survey

Only 13.3% of the respondents in the table above who defined the level of digitalization as low, declare that environmental and social practices are applied in their organizations. In organizations with *good and very good* degree of digitalization, this percentage increases to 36.5%. The ratio in relation to the application of "green" practices is similar - 25% and 55.7% (*Low/ Good and very good* degree of digitalization). A similar proportional dependence is also found in companies that have not implemented additional training courses for their employees and the application of "green" practices - 31.5% and 57.7%. This gives reason to conclude that continuing professional training and digitalization of activities in organizations are factors that positively influence environmental and social practices related to sustainable development and corporate responsibility.

CONCLUSION

Digitalization requires substantial changes in the infrastructure, functionality, and communication practices of organizations. The use of new digital technologies also entails risks that must be considered during the implementation of digital transformation. Digital technologies make it possible to optimize resource use, reduce energy consumption, monitor and analyze environmental data, measure emissions, and improve forecasting and planning. Traditional sectoral approaches are no longer adequate due to the complexity of sustainable development challenges. In many companies, economic efficiency is still evaluated only in terms of costs, resources, and outputs, while environmental and social variables are often neglected. This represents a weakness in systems for assessing economic performance.

Sustainable development requires effective interaction among its three subsystems—economic, environmental, and social. The limited results achieved so far are related to insufficient responsibility among political, managerial, and technological systems, as well as the lack of clear evaluation criteria. Human capital, education, and modern management practices play a crucial role in achieving sustainability. In knowledge-based economies, innovation and creativity are key intangible resources and drivers of long-term value.

REFERENCES

Brennen, J. S., & Kreiss, D. (2014). Digitalization and digitization. *Culture Digitally*. 8(2), 10-19. https://culturedigitally.org/2014/09/digitalization-and-digitization/

Brennen, J. S., & Kreiss, D. (2016). Digitalization. In *The International Encyclopedia of Communication Theory and Philosophy*. 1-11. DOI: 10.1002/9781118766804.wbiect111

European Commission. (2015). *Sustainable Development Goals*. https://international-partnerships.ec.europa.eu/policies/sustainable-development-goals_en

European Commission. (2019). *European Green Deal. Striving to be the first climate-neutral continent*. https://commission.europa.eu/strategy-and-policy/priorities-2019-2024/european-green-deal_en

European Commission. (2020). *Recovery plan for Europe*. https://commission.europa.eu/strategy-and-policy/recovery-plan-europe_en

European Commission. (2021a). *Delivering the European Green Deal*. https://commission.europa.eu/strategy-and-policy/priorities-2019-2024/european-green-deal/delivering-european-green-deal_en

European Commission. (2021b). *European Climate Law*. https://climate.ec.europa.eu/eu-action/european-climate-law_en

European Commission. (2023). *Communication from the Commission to the European Parliament, the European Council, the Council, the European Economic and Social Committee and the Committee of the Regions. A Green Deal Industrial Plan for the Net-Zero Age*. https://eur-lex.europa.eu/legal-content/EN/TXT/PDF/?uri=CELEX:52023DC0062

European Commission. (2024). *Shaping Europe's digital future. DESI indicators*. https://digital-decade-desi.digital-strategy.ec.europa.eu/datasets/desi/charts/desi-indicators?period=desi_2024&indicator=desi_ai&breakdown=ent_all_xfin&unit=pc_ent&country=AT,BE,BG,HR,CY,CZ,DK,EE,EU,FI,FR,DE,EL,HU,IE,IT,LV,LT,LU,MT,NL,PL,PT,RO,SK,SI,ES,S

European Court of Auditors. (2023). *Special report: Circular Economy - Slow Transition by Member States despite EU action.* https://www.eca.europa.eu/en/publications/sr-2023-17

European parliament. (2021). *Report on artificial intelligence in criminal law and its use by the police and judicial authorities in criminal matters.* https://www.europarl.europa.eu/doceo/document/A-9-2021-0232_EN.html

Eurostat. (2024a). *Circular material use rate.* DOI: 10.2908/CEI_SRM030

Eurostat. (2024b). *Consumption footprint.* DOI: 10.2908/CEI_GSR010

Eurostat. (2024c). Greenhouse gases emissions from production activities.

Eurostat. (2024d). *Persons employed in circular economy sectors.* DOI: 10.2908/CEI_CIE011

Eurostat. (2024e). *Population on 1 January by age and sex.* DOI: 10.2908/DEMO_PJAN

Eurostat. (2024f). *Waste generation per capita.* DOI: 10.2908/CEI_PC034

Filipova, M., Gardan, I., Bondrea, A., & Gardan, D. (2024). Relationship between organizational culture and corporate sustainability. *Annals of Spiru Haret University. Economic Series, 24*(1), 19–40. DOI: 10.26458/2411

Frenzel, A., Muench, J. C., Bruckner, M., & Veit, D. (2021). Digitization or digitalization?-Toward an understanding of definitions, use and application in IS research. *Amcis 2021 Proceedings, 18.* Montreal. https://aisel.aisnet.org/amcis2021/adv_info_systems_general_track/adv_info_systems_general_track/18

Legner, C., Eymann, T., Hess, T., Matt, C., Böhmann, T., Drews, P., & Ahlemann, F. (2017). Digitalization: Opportunity and challenge for the business and information systems engineering community. *Business & Information Systems Engineering, 59*(4), 301–308. DOI: 10.1007/s12599-017-0484-2

Lyytinen, K., Yoo, Y., & Boland, R. J.Jr. (2016). Digital product innovation within four classes of innovation networks. *Information Systems Journal, 26*(1), 47–75. DOI: 10.1111/isj.12093

National Programme. "Digital Bulgaria 2025". (2019). https://www.mtc.government.bg/en/category/85/national-program-digital-bulgaria-2025-and-road-map-its-implementation-are-adopted-cm-decision-no73005-12-2019

Ross, J. (2017). Don't confuse digital with digitization. *MIT Sloan Management Review*. https://sloanreview.mit.edu/article/dont-confuse-digital-with-digitization/

Sandberg, J., Holmstrom, H., & Lyytinen, K. (2020). Digitization and phase transitions in platform organizing logics: Evidence from the Process Automation Industry. *Management Information Systems Quarterly*, *44*(1), 129–153. DOI: 10.25300/MISQ/2020/14520

SAP. (2016). Digital business modeling: A structural approach toward digital transformation. *White Paper Digital Transformation.* DOI: 10.13140/RG.2.1.3167.3368

Schaltegger, S., Herzig, C., Kleibir, O., & Muller, J. (2002). *Sustainability management in business enterprises: Concepts and instruments for sustainable organization development. The Federal Ministry for the Environment, Nature Conservation and Nuclear Safety (BMU).* Division for Environment and Economy., http://fox.leuphana.de/portal/files/1198766/German_Federal_Ministry_for_the_Environment.pdf

Serafimova, V., & Vasilev, V. (2024). Digital culture as a competitive advantage in the sustainable development of organizations. *Agora International Journal of Economical Sciences.*, *18*(1), 210–222. DOI: 10.15837/aijes.v18i1.6724

Stefanova, D. P., Vasilev, V. P., & Efremovski, I. P. (2023). Re-innovative organizational design: Sustainable branding and effective communication - applied models in a world with new borders/without borders. In *Handbook of Research on Achieving Sustainable Development Goals With Sustainable Marketing*. 112-127. DOI: 10.4018/978-1-6684-8681-8.ch006

Sustainable Development Report. (2023). *The 2023 Sustainable Development Report is out.* https://sdsn.bg/en/the-2023-sustainable-development-report-is-out/

United Nations. (1987). *Report of the World Commission on environment and development: Our Common Future.* http://www.un-documents.net/our-common-future.pdf

Wade, M. (2015). Digital business transformation. A conceptual framework. *Global Center for Digital Business Transformation.* An IMD and Cisco Initiative. https://www.imd.org/uupload/IMD.WebSite/DBT/Digital%20Business%20Transformation%20Framework.pdf

Yıldırım, S., & Vasilev, V. (2024). The future of green collar workers: Green Transition in employment. In Yıldırım, S., Yıldırım, D., Demirtaş, I., & Kandpal, V. (Eds.), *Green Transition Impacts on the Economy* (pp. 1–16). Society, and Environment., DOI: 10.4018/979-8-3693-3985-5.ch001

Yuleva-Chuchulayna, R. (2024). Models for assessment of competitive positions and the competitiveness of companies. *Journal of Management Sciences and Applications*, *2*(2), 232–244. https://www.unwe.bg/doi/JOMSA/2024.2/JOMSA.2024.2.06.pdf. DOI: 10.37075/JOMSA.2024.2.06

. Yuleva-Chuchulayna, R. & Filipova, M. (2023). The future of sustainable Business: The case for a "Green Economy" of enterprises in Bulgaria. *Geopolitical perspectives and technological challenges for sustainable growth in the 21st century, Sciendo*. 841-848. DOI: 10.2478/9788367405546-077

ADDITIONAL READING

Bostancı, S. H., & Yıldırım, S. (2021). Sustainable Communities vs. Climate Refugees: Two Opposite Results of Climate Change. In Popescu, C. (Ed.), *Handbook of Research on Novel Practices and Current Successes in Achieving the Sustainable Development Goals* (pp. 298–319). IGI Global., DOI: 10.4018/978-1-7998-8426-2.ch015

European parliament (n.d.). Bulgaria's National Recovery and Resilience Plan. Latest State of Play. *Next Generation EU (NGEU) delivery – How are the Member States doing?* https://www.europarl.europa.eu/RegData/etudes/BRIE/2022/733662/EPRS_BRI(2022)733662_EN.pdf

Filipova, M., D. Yaneva, I. Mierlus-Mazilu (2023). Specifics of digital transformation in business. *Economics and management*, 20 (2), 110-121. https://doi.org/DOI: 10.37708/em.swu.v20i2.7

Vasilev, V., Stefanova, D., & Icheva, M. (2024). Green Human Resource Management as a Component of Sustainable Organizational Development in Environmental and Natural Economics. *Environment. Technology. Resources, 1*, стр. 402-407. Veliko Tarnovo. DOI: 10.17770/etr2024vol1.7966

Yildirim, S. (2020). The consumer role for sustainable development: how consumers contribute sustainable development goals. In Chkoniya, V., Madsen, A. O., & Bukhrashvili, P. (Eds.), *Anthropological Approaches to Understanding Consumption Patterns and Consumer Behavior* (pp. 325–341). IGI Global., DOI: 10.4018/978-1-7998-3115-0.ch018

KEY WORDS AND DEFINITIONS

Digitalization: A process of implementation of modern digital technologies and creation of communication and media infrastructures.

Digital Transformation: It is related not only to production processes, but also to marketing, advertising and communication with partners and customers.

Green Digitalization: An analogue of the concept of green economy and an up-to-date economic and technical approach to implementing the ecological aspects of sustainable development.

Sustainable Development: A concept that is related to global, environmental, social and economic problems with a projection into the future.

Chapter 7
Risk Society and Environmental Justice:
A Comparative Evaluation in the Context of Social Work Through the Cases of Turkey and Japan

Esra Soğancı
https://orcid.org/0000-0002-9224-578X
Vocational School of Social Sciences, Tekirdağ Namık Kemal University, Turkey

ABSTRACT

This study explores how flood disasters reflect environmental injustice within the framework of risk society theory and social work. Through a comparative analysis of the 2018 Japan Floods and the 2021 Western Black Sea Flood in Turkey, it examines how modern risks disproportionately affect vulnerable populations and how disaster responses differ based on national resilience, institutional capacity, and social justice. Drawing from Ulrich Beck's risk society and Bullard's environmental justice approaches, the study reveals Japan's reflexive adaptation to disasters, emphasizing prevention, community resilience, and psychosocial recovery, while high-

DOI: 10.4018/979-8-3373-9685-9.ch007

lighting challenges in Turkey's risk preparedness, public participation, and equity in recovery. The research underscores the need for inclusive, rights-based, and socially just disaster management systems.

INTRODUCTION

This study presents a comparative evaluation of the concept of risk in late modern societies within the context of environmental justice, based on two flood disasters that occurred in Japan and Turkey. The choice of flooding as the subject of analysis is based on the fact that it emerges as a manifestation of environmental injustice and disproportionately affects vulnerable population groups, thereby exacerbating environmental inequality. The selection of Japan and Turkey for this comparative analysis stems from their high levels of geological, meteorological, and topographical risk, necessitating an adaptation to a way of life shaped by disaster awareness. Although both countries have strong infrastructure in terms of disaster preparedness, Japan's disaster management system positions it as one of the most prepared nations in the face of natural disasters (Des Marais et al., 2012). In terms of disaster resilience, Japan is significantly more advanced compared to Turkey, and also ranks higher in terms of social justice. In this context, the study evaluates the impact of disasters on environmental justice by focusing on the '2021 Western Black Sea Flood Disaster' in Turkey and the '2018 Japan Floods.' The disaster management systems of Turkey and Japan, along with relevant literature on the cases, are examined through the lens of Ulrich Beck's Risk Society theory and Bullard's Environmental Justice approach within the scope of social work.

Ulrich Beck's Risk Society Theory

Approaching nature as a tool of utility and subjecting natural resources to control and management for the benefit of human life is one of the defining characteristics of modern society. As human power and control over nature increases, environmental crises and new risk factors emerge. On the other hand, as Erickson (2018) states, the way humans and nature influence each other, and the impact of our choices as individuals or communities on the world we inhabit, is not sufficiently understood by many. In this context, Ulrich Beck offers a significant conceptualization with his theory of the risk society. In his 1986 work Risk Society: Towards a New Modernity, Beck (1992) developed the concepts of risk society and reflexive modernization, discussing the structure of late modernity as one that increases contingency and risks.

According to Beck, many ecological risks have emerged in modern societies as a result of industrialization. The crises caused by these risks shape the relationship between humans and nature. While industrial societies develop strategies to protect themselves from hazards through risk assessments, the risks and long-term outcomes of ecological degradation cannot be fully known. Therefore, living in the face of unpredictable and unpreventable risks becomes a characteristic feature of modern society (Stevenson, 2017).

In Beck's view (1992), in late modernity, the production of wealth also leads to the production of social risks. The dangers and risks generated as a by - product of modernization pose a threat to the process itself, making it crucial to prevent or minimize these dangers. Attempting to keep such risks within a tolerable range requires dealing with the problems caused by technological and economic development. This situation renders the modernization process reflexive in characteristics. Risks generated by technological advances necessitate the evaluation of technology not only for its benefits but also for its potential hazards, thus compelling modernization to become self - critical. Therefore, a self-critical society can create a new order. For instance climate change has the potential to inspire systemic change (Beck, 2010a).

Although danger has always existed in human life, compared to pre - industrial times, the way modern societies experience and perceive risks is significantly different. This difference is less about the frequency or magnitude of risks and more about how they are perceived. In pre - industrial societies, risks were often seen as fate —beyond human control— whereas in industrial societies, they are approached as calculable decisions through the concept of 'risk notion.' In this framework, hazards can be foreseen, risks can be calculated, prevented, or insured against (Stevenson, 2017). Furthermore, the risks caused by advanced modernization differ greatly in scope and impact. Industrialization related risks, such as climate change, threaten not just specific areas but global life as a whole. Even countries with no role in creating these hazards may suffer their consequences, clearly illustrating this disparity. As Beck (2010b) said; global risk has two sides: the traumatic vulnerability and the resulting responsibility for all, including one's own survival." Another dimension of differentiation involves the visibility of risks. Dangers that were clearly observable in pre - modern societies are not as easily detected today. Identifying and understanding many risks requires technical and scientific expertise. For examples, a technological manipulation and scientific understanding is required to identify levels of hazardous chemicals in drinking water or develop models of climate change (Cordner, 2021).

For all these reasons, today's dangers and risks differ significantly from those faced by pre -modern societies. The evolving nature of dangers and risks created by modernization necessitates a new type of society that learns to live with these risks. Beck describes this situation using the concept of 'reflexive modernization.' As new dangers are perceived, modern society becomes more self - critical. This is because addressing the new kinds of risks brought about by technological and economic development requires a critical framework (Beck, 1992). In other words, modern risks bring with them a new type of society that demands contemplation, expertise, knowledge - based systems, and self - criticism (Turner, 2006). Risks that are not easily detectable only become political issues once they are recognized by people. Therefore, scientific explanations, professional opinions, and mass communication tools are needed to define risks (Beck, 1992). Laws specific to the issue, developed technologies, expert opinions and guidance are employed to manage risks. However, scientific and economic development also produce risks that cannot be prevented or insured against, revealing the limitations of reflexive society. Therefore modern societies have embrace their improved, though ultimately incomplete, ability to understand, calculate, and predict future risks produced by scientific and economic systems (Cordner, 2021). For this reason, all established norms and institutions — risk assessments, insurance systems, and preventive measures — may fail in the face of new dangers (Beck, 1992). Moreover, actions to manage the risks of modern society are often controlled by those in power and wealth, which makes them controversial (Stevenson, 2017).

Risks that are unleashed through laws and regulations to maintain economic growth end up harming some people. Particularly, the poor and vulnerable are disproportionately affected by the distribution and escalation of risks. Nevertheless, the risks we face also threaten those who produce and profit from them (Beck, 1992).

Disasters in Risk Societies

Disasters, which are directly linked to human actions and inactions, are among the fundamental risks created by modern society and at the same time, they are challenges it seeks to manage. By definition, disasters are not the natural events themselves but the destructive outcomes resulting from insufficient preparedness against such events. Thus, disasters are more about the capacity of infrastructure to adapt to hazards rather than the natural events per se (Lin et al., 2020). Modern society continually generates new strategies with its evolving knowledge, technology, legislation, and policies to prevent events from turning into disasters. Although predicting disasters is difficult, global examples show that significant levels of risk can be mitigated. However, human interventions in nature via science and technology introduce human-made elements into the creation of disasters. While various activities that enable and facilitate human life continue, they also increase environmental degradation, thereby exacerbating disaster risk (Görgün Baran, 2023).

Whether an event is defined as a disaster depends on its ability to overwhelm local coping mechanisms and disrupt daily routines. Coping capacity is the combination of strengths, attributes, and resources that a community, society, or organization possesses to manage adverse effects. One key determinant of this capacity is vulnerability, defined as exposure to damaging external impacts (Oxfam, 2011). All societies have different levels of vulnerability. Since 2012, Index for Risk Management (INFORM) has evaluated countries' disaster risks based on dimensions such as 'hazard and exposure', 'vulnerability', and 'coping capacity'. According to the 2025 risk index, Africa contains 11 of the 16 highest-risk countries, making it the most at-risk region (INFORM, 2025).

Disasters, as defined globally by The International Disaster Database (EM-DAT), are natural events that result in at least one of the following:

10 or more fatalities, 100 or more people affected, declaration of a state of emergency, or an international appeal for assistance (EM-DAT, 2024b). Although living spaces are inherently exposed to various geographical hazards and risks, the potential for these to escalate into disasters can be mitigated through individual, institutional, and structural measures. A country's available resources influence the disaster risk reduction strategies it can develop; however, due to the unpredictable and transboundary nature of modern risks, absolute preparedness is unattainable. Therefore, as Dominelli (2015) states, even the most disaster-resilient countries can be affected by disasters because, in the modern world, it is impossible to predict or prepare for every risk.

Climate change, as a negative side effect of industrialization, has become one of the primary unmanageable risks in today's world. Meteorological disasters triggered by climate change are increasing in both frequency and severity. Developed countries, as major emitters, significantly contribute to global warming. The Intergovernmental Panel on Climate Change (IPCC, 2007) recognizes climate change caused by global warming as one of the primary global threats.

In order to eliminate this risk, international agreements impose various obligations on states, yet the failure of states to fulfill these obligations increases the risks faced (Dominelli, 2024, Directorate of Climate Change, 2024). The continuous increase in the destruction caused by disasters on human life globally (EM-DAT 2024a) reveals that the risks we face are growing and that modern society is inadequate in coping with them.In this context, the 2022 Global Natural Disaster Assessment Report (UNDRR, 2023) and EM-DAT (2023) data show that more than half of the natural disasters occurring in the world are meteorological in nature, and the impact and frequency of meteorological disasters are increasing each year. Floods, droughts, heatwaves, and forest fires linked to climate change are causing significant destruction across the globe. Among these, floods are expected to become an even greater threat in the future, resulting in considerable disaster damage each year (Dominelli, 2024). Especially coastal areas, including the densely populated mega-delta regions of South, East, and Southeast Asia, are increasingly facing the threat of flooding with greater intensity (Intergovernmental Panel on Climate Change, 2007).

The loss of life and property, injuries and illnesses, psychological disorders, and stress-related conditions resulting from disasters cause structural changes in the affected communities and even in the broader society (Yalçın, 2020). Disasters are thus acute, collective, and traumatic events (McFarlane & Norris, 2006). Human-induced environmental degradation —especially when approached from an economic growth-oriented perspective— exacerbates the negative impacts of disasters on individuals and habitats. Urbanization, population growth, settlement in unsuitable areas, vegetation destruction, and concrete expansion are among the contributing factors (Dominelli, 2024).

However, even in regions prone to hazards, if infrastructure is robust and adaptive (Lin et al., 2020), engineering solutions such as early warning systems and risk-aware land-use planning can be implemented. When social protection and funding mechanisms are made accessible to all (Dominelli, 2024), the transformation of hazardous events into disasters can be prevented. For this reason, within the framework of the risk society, it is necessary to recognize hazards and interrogate the human activities contributing to their emergence. At the same time, structural, institutional, and human - centered measures should be continuously developed to prevent recognized risks from escalating into disasters. A risk management system that accounts for social context and enhances the resilience of vulnerable population groups is key to achieving environmental justice.

Environmental Justice

Studies related to environmental justice coincide with the 1960s in the United States when environmental movements achieved legislative goals, and the protests regarding environmental injustice in the 1980s (Maung and Pellow, 2021). The book written by Bullard in the 1990s titled 'Dumping in Dixie: Race, Class, and Environmental Quality,' significant for being the first book written on environmental justice, places environmental justice discussions on an academic foundation. This work, based on the assertion that all people, regardless of differences, have the fundamental rights to live, work, receive education, play, and worship in a healthy environment, examines economic, health, and environmental inequalities through various case studies (Bullard, 2000). Within these case studies,

Bullard investigates the areas in the United States where harmful facilities such as garbage collection sites and hazardous waste plants are located, and determines that regardless of whether the residents are wealthy, middle-class, or poor, these facilities are predominantly established near African American communities (Bullard, 2000). As a result of this investigation, academic interest emerged in the idea that environmental hazards are disproportionately experienced by certain population groups and that this phenomenon is not solely based on socioeconomic status.

Environmental justice is defined as an interdisciplinary academic field focusing on the unequal experiences of environmental harm and benefits among population groups, as well as a grassroots social movement (Maung and Pellow, 2021). Environmental justice is the equal sharing of natural resources and environmental damages. It is a concept that refers to the right of all people, regardless of categorical differences such as income, race, ethnicity, national origin, or disability, to be treated fairly in the face of environmental impacts and to participate in decision-making mechanisms that affect human health and the environment. In this context, ensuring environmental justice means that no individual or group should be disproportionately harmed by environmental hazards, risks, and negative effects (EPA, 2024). According to Bullard (2000), environmental justice is related to poverty but is a political and racial issue that cannot be reduced to that alone. Environmental justice occurs when all people benefit significantly from environmental protection, when no group is excluded from environmental decision-making processes, and when no group is disproportionately affected by environmental harms (Erickson, 2018).

Therefore, debates on environmental justice focus on the disproportionate exposure of people of color, Indigenous peoples, immigrants, and the working class to environmental threats and risks (Bullard & Wright, 1987). Environmental inequalities are closely related to the inequalities experienced by individuals from low socioeconomic classes and racial minorities (Encyclopedia of Philosophy, 2024). Today, the term "environmental justice" has been used to highlight the distributional impacts of the dominant society's environmental decision-making process on disadvantaged communities, including the poor and racial minorities (Tsosie, 2007).

Environmental justice studies fundamentally focus on two issues: the disproportionate impact of environmental risks on disadvantaged population groups and the fact that environmental crises are largely driven by processes that generate social inequalities (Maung and Pellow, 2021).

Environmental risks disproportionately affect disadvantaged population groups. Environmental degradation places a heavier burden on areas marked by poverty, poor governance, low rent, and low-status jobs. These regions often face health hazards such as air pollution and industrial waste (Bullard, 2000). Marginalized populations typically reside in such locations, characterized by housing insecurity, lack of services, and disease outbreaks. Due to their categorical disadvantages, individuals in these communities are disproportionately affected by risks due to inadequate infrastructure and regulation (Dominelli, 2024).

Environmental crises are largely driven by processes that generate social inequalities. Although developed countries have historically contributed more to greenhouse gas emissions, climate change is expected to have a greater impact on poorer and developing nations (IPCC, 2007). Developing countries, which occupy a disadvantaged position in the global capitalist system, are more vulnerable to climate crises because they have fewer resources to adapt to climate changes and to protect themselves from adverse effects. This dual threat shows that climate crises are undeniably an issue of environmental justice (Miranda et al., 2011). Moreover, when compared to developed countries, these nations have contributed far less to the emergence of climate crises, yet due to their limited resources, they cannot protect themselves from these changes and are more severely affected (Maung & Pellow, 2021).

While early studies on environmental justice originated in the U.S. and focused on the geographic distribution of toxic environments, today the concept has evolved into a global discourse. Since the 2000s, increased awareness of climate change and the vulnerability of developing countries have expanded the academic and political focus on environmental justice. The inherently unequal nature of climate crises has broadened the environmental justice debate to include topics such as food justice, climate justice, and energy justice (Maung & Pellow, 2021). Hurricane Katrina, which struck the Gulf Coast of the U.S. in 2005, is considered a turning point for framing environmental justice in the context of disasters. After the hurricane, the vulnerabilities of disadvantaged population groups in the face of meteorological disasters became particularly noticeable. Today, the environmental justice framework serves as a vital lens for examining how vulnerable groups are disproportionately affected by disasters. Research shows that these populations are more exposed to natural hazards, suffer greater damage, and take longer to recover (Vilá et al., 2023).

Traumatic disaster events, which create new inequalities and deepen existing ones in the social structure, disproportionately affect low-income, marginalized communities, especially because they live in environmentally sensitive areas. These communities experience significant difficulties not only due to the destruction caused by disasters but also in basic areas such as health and economy, largely because of their lack of preparedness. Dominelli (2014) defines environmental injustice as "the failure to equitably distribute resources to meet human needs and ensure the current and future well-being of both people and the planet." Reversing this condition requires the realization of environmental justice. Achieving environmental justice first requires dismantling structural inequalities (Dauvergne, 2009). A comprehensive approach that includes legal frameworks, geographical conditions, residential distribution, social vulnerabilities, and inclusive participation must be adopted to ensure equity among individuals and social groups (Bullard, 2001).

The prioritization of one group at the expense of others or the protection of its interests poses a significant threat to environmental justice. For this reason, environmental justice inherently requires social justice. Considering the social consequences of environmental disasters, the relationship between environmental justice and social justice is reciprocal. Accordingly, every effort aimed at advancing social justice directly or indirectly contributes to the realization of environmental justice.

The social work discipline, which aims to promote social justice, has, since the 1990s, adopted an "individual-in-environment" approach to protect the physical environment. Green social work emphasizes the intrinsic value of the physical environment and defends nature's right to exist independently. From the perspective of green social work, environmental justice has evolved to align more closely with the demands of the 21st century, gaining a more holistic framework. It recognizes the symbiotic relationship between humans and space and supports everyone's right to a healthy environment. By advocating for the equitable distribution of resources, green social work envisions structural change that addresses various forms of oppression, such as poverty, racism, and sexism (Dominelli, 2024). The social work discipline, committed to the pursuit of social justice, plays a critical role in the achievement of environmental justice. This is because the populations that social workers most frequently serve are often the same disadvantaged groups most severely affected by adverse environmental conditions and disasters (Demirbilek, 2016). From this perspective, the causes and consequences of environmental crises are not only linked to the global economy but also to social and political issues such as gender discrimination, colonialism, and racism, necessitating a comprehensive justice approach.

DISASTERS AS BOTH A CAUSE AND CONSEQUENCE OF ENVIRONMENTAL INJUSTICE

The Case of the 2018 Japan Floods: The Case of Japan

Japan is located in the Pacific Ring of Fire, a volcanic zone that accounts for 90% of global earthquakes (Livescience, 2011). Due to its geographical location, Japan faces extremely high disaster risk and ranks as the fourth most disaster-prone country globally (World Bank, 2019). Natural disasters deeply affect both economic and social life, prompting Japan to adopt a philosophy of “living with disasters”. In accordance with this reality, Japan has developed a comprehensive disaster risk management system.

Japan has strong disaster preparedness measures. However, its disaster management system is primarily earthquake - focused. While this approach has made Japan resilient to earthquakes, it has also led to increased vulnerability to floods and other disasters. Similarly, the general public is less aware and prepared for disasters other than earthquakes (World Economic Forum, 2018). Moreover, with two-thirds of its land consisting of mountainous terrain, Japan is geologically exposed to other disaster risks such as landslides and slope failures. Wooden houses, often preferred in residential areas for their earthquake resistance, create vulnerabilities to floods and landslides, especially when built on sloped terrain (Lin et al.,

2020). As a result, Japan continues to experience devastating disasters—such as the one in 2018—due to unforeseen or underestimated risks and inadequate preparedness at both the individual and institutional levels for flood-related events.

As one of the countries most prepared for disasters, the Japanese government manages many disaster events in collaboration with local professionals by focusing on enhancing community resilience and increasing capacity to cope with challenges (Des Marais et al., 2012). Japan has also developed an "Adaptive Social Protection" system to provide social and economic support to disaster victims, including the most vulnerable populations (World Bank, 2019). This system is re-evaluated after each disaster and continuously revised in line with new laws and institutional developments. Based on the recognition that the most disadvantaged groups are also the most affected by disasters, Japan's Disaster Risk Management and Adaptive Social Protection systems place the social support needs of vulnerable populations at the center of intervention strategies (World Bank, 2019). Ensuring public participation is one of the most critical aspects of the disaster management system (Des Marais et al., 2012). In Japan, civil society organizations and volunteers are involved in all phases of disaster management, from emergency response to long-term recovery. Volunteers, both those previously organized and those mobilized immediately after the disaster, often play a key role in early interventions. Civil society organizations from various parts of the country coordinate their response in close collaboration with local communities and government officials (CWS Japan, 2023).

Although disaster resilience is a core criterion in everyday life and urban planning in Japan, large - scale losses still occur despite extensive protective and preventive efforts. After such events, Japan acknowledges the shortcomings of its existing systems and continuously strives to enhance preparedness and reduce disaster risks. These ongoing efforts serve to further improve Japan's disaster management framework, turning it into a global model. In this context, the 2018 Japan Floods are examined as a case study in which Japan both experienced a major disaster and used the lessons learned to further develop its disaster management practices.

In July 2018, catastrophic floods occurred in 31 prefectures across western Japan. Data from 123 observation stations revealed that the torrential rainfall triggered by Typhoon Prapiroon reached record-breaking levels. As a result, water levels rose more than three meters within a very short period (Lin et al., 2020), making this disaster the most significant flood in Japan since 1982. The disaster caused major damage: 237 people lost their lives, 8 were reported missing, 433 were injured, and 6,767 homes were completely destroyed. The damage caused by the floods was estimated at approximately 9.86 billion USD (Yoshida, et al., 2023). The most severely affected areas were the prefectures of Hiroshima, Okayama, and Ehime. Due to extensive infrastructure damage, electricity and water supplies were disrupted, and two million people were evacuated from the affected regions (Miyamori et al., 2023). The damage to the transportation system hindered public services and delayed emergency interventions (Yoshida, et al., 2023). In addition to the floods, several secondary disasters such as landslides, collapses, and mudflows occurred. Due to their combined impact, this flood was the first in Japan to be officially declared a "Special Emergency Disaster – Catastrophic Disaster." The event exposed the inadequacy of Japan's flood prevention and risk management systems (Lin et al., 2020).

Post-disaster interventions are significantly influenced by the strength of systems already present in society prior to the disaster. In this context, it is important to note that Japan has a robust social protection program designed to safeguard disadvantaged populations, support them, and guarantee minimum living standards during normal times. This program is a security system that focuses on specific vulnerable population groups such as single parents, people with disabilities, and families with young children. In the event of a disaster, various emergency assistance programs and additional supports are implemented alongside the regular social protection scheme. These include cash transfers covering needs such as basic necessities, loss of family members, home repairs, and injuries; fee waivers for poor households; and interest-free loan options (World Bank, 2019). In addition to official aid and support, donations are distributed through local governments, and programs such as "Cash for Work" and unemployment insurance are activated in the aftermath of the disaster. Long-term assistance is provided to affected individuals under the "Law Concerning Special Fiscal Aid for Coping with Disasters in Japan" (Miyamori et al., 2023).

Following the 2018 flood disaster, affected individuals were classified and certified by local governments based on the level of impact they had experienced. This allowed disaster assistance to be planned according to need, rather than offering uniform interventions (Yoshida et al., 2023). According to the Japanese Red Cross Society, medical emergency, coordination, and relief teams were deployed in the disaster area. Psychosocial support teams actively conducted needs assessments, distributed emergency kits and essential supplies, managed donation processes, and ensured the effective organization of volunteer activities (Japanese Red Cross Society, 2018). Nearly 4,000 shelters were established after the disaster and remained operational until December of the same year. Dozens of high-level meetings, including those attended by the Prime Minister, were held to make decisions regarding damage assessment, response coordination, and the prevention of secondary disasters (Disaster Management in Japan, 2022).

There is a strong culture of volunteerism and mutual aid in the country. After the disaster, volunteers and civil society organizations played crucial roles in meeting both the physical and psychological needs of those in evacuation shelters (World Bank, 2019). In line with this culture, numerous "Disaster Volunteer Centers" were established in flood-affected regions. Around 260,000 volunteers from across Japan arrived to directly assist disaster victims by cleaning mud from homes and helping reorganize living spaces. To prevent social isolation due to displacement and to ensure continued access to social opportunities, the central government provided care and counseling services (Disaster Management in Japan, 2022). Moreover, "social care services" contributed to enhancing social resilience by enabling and supporting the participation of disadvantaged population groups (World Bank, 2019).

The lessons learned and the interventions undertaken after the disaster were systematically documented with the goal of using this painful experience to strengthen Japan's disaster management system. In this regard, in Mabi Town, part of Okayama Prefecture and one of the areas most severely affected by the 2018 Japan Floods, post-disaster interventions were regularly evaluated through direct observation, interviews, and document analysis conducted every six months and again five years after the disaster. These evaluations covered various aspects such as early warning systems, risk communication, infrastructure management, evacuation procedures, shelter operations, and volunteer coordination (CWS Japan, 2023). Mabi Town experienced extensive damage due to its location between two rivers that overflowed during the flood. When the levees broke, water levels rose up to five meters in many areas and affected approximately 4,600 homes. Fifty-one people—most of whom were over the age of 70—lost their lives (CWS Japan, 2019). Due to the flood and subsequent evacuations, many survivors experienced health problems, and vulnerable population groups such as the elderly and disabled who required special dietary restrictions or continuous medication, as well as families with young children, were disproportionately affected. In addition, a significant number of survivors experienced psychosocial stress (CWS Japan, 2019).

Five years after the disaster, nearly all of the damaged houses had been rebuilt, and several interventions aimed at enhancing disaster resilience were implemented. These included the restoration and reinforcement of levees, the widening of the riverbed, and the reorganization of river channels in city centers to reduce the risk of urban flooding. Additionally, new facilities such as a "disaster prevention park" and a "river disaster prevention station" were constructed along the riverbanks to ensure that the rivers—once sources of catastrophic floods—could continue to serve as symbols of strength and inspiration for the local population (CWS Japan, 2023). Using sediment and debris carried by the floodwaters, new multifunctional spaces were created. These areas are intended for sports, recreation, and disaster awareness activities during normal times, and as gathering, evacuation, shelter, and storage spaces during emergencies. Local authorities emphasized that the interventions aimed for more than mere restoration; they sought to create stronger, more resilient living environments, and were committed to establishing a system in which no disaster survivor would be overlooked (CWS Japan, 2023).

To support psychosocial health and promote social cohesion, civil society organizations organized "salons"—informal event and meeting spaces—for individuals who had been displaced from their homes due to the disaster. These venues took various forms, from shared spaces for information exchange and mutual emotional support, to communal areas where people could cook and eat together. The opportunity for affected communities to come together played a significant role in maintaining psychosocial recovery and social integration (CWS Japan, 2019).

In addition to enhancing the city's resilience, considerable attention was given to raising awareness among local residents in preparation for potential future disasters. Educational efforts included the distribution of printed materials, the launch of informative websites, the sharing of risk maps, marking flood water heights on buildings, and indicating the maximum water levels that levees could hold. Furthermore, early warning systems and evacuation plans were improved based on the experiences of disaster-affected individuals, and the updated guidelines were made publicly available (CWS Japan, 2023).

The 2021 Western Black Sea Flood Disaster: The Case of Turkey

Natural disasters frequently occur across Turkey's geography, causing significant damage to individuals' life and property security. In this context, not only structural factors that contribute to the occurrence of disasters but also social factors —such as the lack of adequate individual preparedness — and organizational factors — such as institutional inability to demonstrate full readiness— are influential.

When examining Turkey's disaster management system, it is observed that until 2009, post - disaster interventions were carried out by separate units and institutions acting independently. However, the experiences gained from various disasters highlighted the need for a continuously operating and coordinated unit. In response to this necessity, the Disaster and Emergency Management Authority (AFAD) was established in 2009 as an institution responsible for centralizing authority and accountability in disaster-related matters and ensuring interinstitutional cooperation (AFAD, 2018). By 2013, with the adoption of the Turkey Disaster Response Plan (TAMP), service groups comprising ministries, public institutions, civil society organizations, and private individuals were defined and assigned responsibilities in responding to all types of disasters (AFAD, 2018). Hence, Turkey's disaster response framework shifted from a crisis management approach to one centered on risk management. Today, it is possible to talk about a system that involves identifying hazards and risks in advance, implementing necessary precautions, coordinating effective responses, and conducting post-disaster recovery efforts in an integrated manner (AFAD, 2022).

Nevertheless, despite all precautions, the Western Black Sea Flood disaster occurred in August 2021. The disaster affected seven districts within the provinces of Kastamonu, Bartın, and Sinop, with Bozkurt district of Kastamonu suffering the most severe damage. The flood, which rendered approximately 80% of the district center unusable (Yeşilyurt, 2023), occurred due to the widespread impact of rains that began on August 11, 2021. While erosion occurred in the upper parts of the district, debris carried by water masses in the lower parts caused extensive destruction to residential zones, agricultural fields, transportation structures, and tourism, commercial, and industrial areas (Bilgen, Balcı, Kalça, 2022). The high level of disaster damage is attributed in part to the district's location within a mountainous floodplain (Karagüzel et al., 2021) and its steep and rugged terrain (Bozkurt Municipality, 2010). In other words, the region is geographically vulnerable to risk. However, additional human-induced factors also contributed to the increased flood risk. These include unauthorized and technically inconsistent interventions in the riverbed running through the district center, the opening of the riverbed area for settlement, and the misplanning of bridge piers in terms of width and height. For these reasons, ponding occurred during heavy rainfall, and in certain areas, the water level rose up to 5.5 meters (Yeşilyurt, 2023, Karagüzel et al., 2021). These observations indicate that the flood was caused by a combination of the region's climatic characteristics, geographical structure, and anthropogenic factors. Nevertheless, field investigations suggest that human activities played a more significant role in the disaster than natural causes (Bilgen, Balcı, Kalça, 2022). In other words, while a portion of the flood damage was due to natural forces, a considerable part stemmed from the failure to align human social practices with natural conditions (Akbal, 2023).

As a result of the disaster, 72 individuals lost their lives, homes, workplaces, roads, and bridges were destroyed, and urban infrastructure was severely damaged. Residents who sought refuge on rooftops to escape the floodwaters found themselves stranded when the water levels receded. Furthermore, the disaster caused damage to public buildings, leading to the loss of documents, information, and equipment essential for emergency response. Public servants who were expected to conduct the initial interventions were themselves turned into victims of the disaster (Soğancı, 2024). In conclusion, the transformation of a natural event such as heavy rainfall into a disaster that disrupts daily life can largely be attributed to human and institutional factors.

Following the disaster, several essential interventions—such as ensuring safe areas, meeting emergency and basic needs, providing psychosocial support, establishing information centers, and protecting public health—were carried out under the coordination of AFAD through public institutions, local authorities, and civil society organizations. Search and rescue operations began immediately, and more than half of the district's population was evacuated from the affected area (AFAD, 2021). Within the scope of emergency response efforts, basic needs such as food, clothing, bedding, and household items were supplied (ASHB, 2022). In addition to in-kind donations, appropriations from the ministry's budget were allocated to address the flood disaster. In this disaster, which damaged the household belongings of 548 families (ASHB, 2021b), civil society organizations, particularly the Red Crescent, along with volunteers, assumed significant responsibilities. For instance, according to the announcement dated January 25, 2021, the Red Crescent provided household appliances and furniture to 113 families based on identified needs (Türk Kızılayı, 2022).

Since the local food production process was disrupted due to the flood, civil society organizations were required to take responsibility in reactivating local resources and supplying food to the region. Containers were set up to address the need for shelter and to allow flood-damaged businesses to resume operations. Grocers, hairdressers, barbers, and coffeehouses continued their activities for a time in these containers. Health-related interventions were provided by healthcare professionals who were temporarily assigned or volunteered during the disaster. Additionally, as all three pharmacies in the district were damaged, pharmaceutical needs during the acute phase were met through mobile pharmacies (Olağandışı Durumlarda Sağlık Hizmetleri, 2021). Psychosocial support services commenced immediately after the disaster (ASHB, 2021a), with intensive support delivered during the first two months and follow-up activities continuing over the following four months—lasting about six months in total. Considering the number of individuals benefiting from psychosocial services and the total district population, it can be stated that these services reached nearly all those affected by the disaster (Soğancı, 2024).

According to an evaluation conducted two years after the flood disaster, the presence of a support model that accounted for all needs, the perceived responsiveness of this model by those who experienced the disaster, the speed of reconstruction efforts, the presence of psychosocial support teams in the field, and the intervention style that instilled a sense of trust collectively contributed to emotional relief. The negative effects of the disaster on individuals and communities were alleviated by the restoration of a sense of trust (Soğancı, 2024). Similarly, findings from another study focusing on the 2021 Western Black Sea flood in the Ayancık district showed that individuals who perceived post-disaster interventions as inadequate and lacked trust in official support experienced more intense levels of traumatic stress (Pak, 2023).

Nevertheless, several shortcomings were observed, including the lack of effective collaboration and coordination between institutions, the fact that some central coordination units became disaster victims themselves, the exclusion of local community opinions in decision-making processes, and the advantage gained by pre-disaster socioeconomically privileged individuals in the allocation of support and compensation. Additionally, the failure to plan interventions based on the specific needs of all affected groups and the inability to ensure equity and justice in the distribution of disaster housing were also noted (Soğancı, 2024). Similarly, Pak (2023) emphasized that vulnerable groups existed before the disaster, that there was insufficient public awareness and information about services, that support and aid were not equitably disseminated, that the reconstruction process was not aligned with local priorities and preferences, and that pre-disaster preparedness and risk reduction systems were found to be inadequate. These negative experiences collectively point to an increase in environmental injustice following the disaster and reveal how such injustice becomes further entrenched as a manifestation of broader social inequality.

CONCLUSION

Disasters and global challenges such as climate change, which adversely affect human well-being across the world, invite the social work profession to play a more active role in addressing the inequalities these phenomena create (IFSW, 2020). The contributions of the social work profession to disaster work are considerable. Via holistic and people-centered approach, social workers recognize special needs of vulnerable populations -like child welfare and elder care issues- that others overlook (Thomas & Healy, 2010). The approach of social work, which considers the individual and their environment holistically, creates a responsibility to engage in regulating human-environment relationships and necessitates the application of specialized knowledge, values, and skills to help individuals overcome environmental challenges and social and economic inequalities. From the very inception of the profession, Jane Addams emphasized the importance of enhancing the living environment with social facilities for the overall welfare of society. For this reason, environmental justice has always been a relevant issue for the science and practice of social work, which centers its interventions on social justice. In recent decades, this issue has matured into a more visible and structured field of study (Erickson, 2018).

From a human rights - based perspective, the goal of interventions in environmental disasters such as floods is to enhance the capacity of individuals to cope with adversity. In line with this objective, the discipline of social work seeks to empower service users, hold service providers accountable, and advocate for risk reduction efforts (Dominelli, 2012). In this context, states are expected to assume responsibilities to ensure the well-being of their citizens—such as preventing disasters, warning citizens about potential risks, implementing emergency responses, engaging in reconstruction efforts, and collaborating with affected populations throughout all these processes (Dominelli, 2024). While many of these responsibilities pertain to acute and long-term interventions after a disaster has occurred, others are related to actions taken beforehand. These include eliminating existing social and environmental injustices in society and ensuring equality among all individuals and social groups in the face of environmental opportunities and threats. Therefore, when disasters are addressed from a human rights perspective, the primary objective should be the realization of environmental justice in the face of risks generated by the risk society. In this view, it is vital to regard disasters as guidance for building a social order in which no one is left behind. As emphasized in the Sendai Framework, post-disaster long-term efforts should not merely focus on replacement of what was lost, but on building something more equitable and improved (United Nations, 2015). In this regard, this section of the study offers an evaluation of the dangers associated with risk society, environmental injustice, and disaster experiences through selected case studies within the context of social work.

In conducting this evaluation, it is important to first mention the Index for Risk Management, which has evaluated and ranked countries based on their risk levels since 2012 using the dimensions of "hazard and exposure," "vulnerability," and "coping capacity." According to the 2025 data, Japan, with a risk level of 2.3, is in the low-risk group, while Turkey, with a risk score of 4.9, falls within the medium-risk group (INFORM, 2025). In the "lack of coping capacity" dimension, Japan scores 1.4, whereas Turkey scores 3.2. These figures indicate that despite facing a higher number of natural hazards geographically (8.3), Japan is subject to lower risk than Turkey (5.5), largely due to its established approach to living with disasters.

Furthermore, according to the United Nations Human Development Index (HDI) 2023/2024 data, which ranks countries based on indicators in education, economy, and health, Japan ranks 24th globally with a score of 0.920, while Turkey ranks 45th with a score of 0.855 (UN, March 13, 2024). Social justice index data for 41 EU and OECD countries shows that Japan ranks 27th, while Turkey ranks 40th (Euronews, 2021). These index data reveal that Japan holds a more advantageous position than Turkey in terms of social welfare and social justice. This is highly significant, as a state's level of preparedness and the negative experiences endured by its citizens during disasters are closely influenced by such factors.

On the other hand, emissions and the resulting global climate change originate from historically industrialized and developed countries such as China, the United States, India, Russia, and Japan. Due to its level of development, Japan is both a contributor to the creation of modern risks and a nation that directly experiences their consequences through extreme weather events. In response to these modern societal risks, Japan employs various strategies such as expertise, knowledge-based systems, scientific research, technology, legal and political regulations. These include not only mechanisms for coping, such as disaster response systems and resilience-enhancing measures, but also broader climate action plans aimed at reducing global carbon emissions and mitigating the effects of the climate crisis. However, as Stevenson (2017) argues, scientific and economic advancement also gives rise to unavoidable risks, leading to circumstances in which the reflexive society proves inadequate. Disasters occur precisely where all established norms and institutions fail. In this regard, the Japan case illustrates how a devastating catastrophe can serve as a milestone for a transformation in mindset. As required by integrated disaster management, the recovery process is regarded as the beginning of preparedness for the next disaster, and the lessons learned from the current one are immediately utilized to build awareness and readiness before they are forgotten (CWS Japan, 2023).

In this context, the efforts undertaken after the 2018 Japan Disasters to create safer living environments and to strengthen disaster resilience—both through structural and institutional measures and public awareness campaigns—are particularly significant. An intervention strategy was adopted that not only strengthened the connection between local people and their environment, but also prevented the hazards inherent in that environment from being forgotten. Examples of this include disaster parks and museums, as well as systematic exposure techniques such as high-water markers incorporated into post-disaster reconstruction. In contrast, such a philosophy of "living with disasters" is not observed in the case of the 2021 Western Black Sea floods in Turkey (Soğancı, 2024). It is necessary for central and local governments in Turkey to update urban planning practices and engage in comprehensive risk reduction initiatives. Furthermore, in order to increase public awareness and knowledge of flood risks, efforts must be made at the household level in areas such as disaster response, risk perception, protective household modifications, emergency evacuation, and the use of protective equipment (Pak, 2023).

As Karataş (2023) notes, what is done after a disaster is directly related to what was prepared beforehand. In Japan, there is a well-structured welfare program in place that provides support for vulnerable populations even in times of normalcy. In times of disaster, assistance is similarly customized, taking into account both needs and disaster-related losses. Evaluative studies in Turkey, however, show that social aid efforts have been insufficient in addressing both pre-disaster conditions and the damage incurred during the disaster. The needs and expectations of disaster-affected individuals vary depending on their situation, stage of life, pre-existing conditions, levels of awareness, and available resources. According to Inter -Agency Standing Comitee (IASC) (2007), an effective psychosocial support service model should focus not only on widespread community losses but also on individual problems that cannot be addressed by existing support and services. In this respect, establishing a needs-based, customized assistance system that evaluates both pre- and post-disaster losses is a critical sensitivity necessary for ensuring social and environmental justice. Additionally, it is important to create new opportunities that enable disaster-affected individuals to maintain connections with their living environments and social networks. To promote social solidarity, disaster survivors should become part of a mutual support mechanism; spaces should be created where they can share emotions and experiences, and new public areas should be designed to facilitate participation in socio-cultural and sports activities. While such urban spaces are present in the Japanese case, in the Turkish context, the reconstruction process has been limited to rebuilding urban areas damaged by the disaster. Moreover, since community participation in decision-making processes during reconstruction has not been adequately ensured, a sense of belonging has not been effectively supported.

Disasters destroy living spaces, exacerbate chronic illnesses, and increase the need for care thus, such traumatic life experiences have a deeper impact on already vulnerable populations. Similarly, Miyamori et al. (2023) emphasize that disasters pose a critical risk factor for elderly individuals by initiating or worsening health problems such as trauma and infection, causing social isolation, disrupting daily activities, and leading to detachment from living environments. Flood disasters also significantly affect mental health. Studies in this field show that mental health conditions such as depression, anxiety, and PTSD increase proportionally with the extent to which individuals lose or suffer damage to their homes due to flooding. For instance, those affected by flooding were found to be six times more likely to suffer from depression, anxiety, and PTSD than those who were not affected (Public Health England, 2020). People who live in flood-damaged areas, have lost a loved one, experienced trauma, or face difficulties in accessing emergency services are more exposed to the cognitive and psychological impacts of disasters. Therefore, a rapid and effective response to disasters is crucial for maintaining psychosocial well-being. When urgent basic needs are not met effectively, psychosocial well-being cannot be achieved either. For this reason, psychosocial support is conceptualized as a pyramid, with essential services and safety-related interventions forming its base (IASC, 2007). In both case studies, it is evident that flood disasters have damaged living spaces and that disadvantaged population groups are more vulnerable to environmental risks during such processes. However, the provision of essential services and safety, along with the timely and effective implementation of acute phase interventions, has contributed positively to the protection of psychosocial health.

The evaluation conducted through the case studies suggests that Japan demonstrates more characteristics of a reflexive society in the face of modern risks and has developed the capacity to live with disasters through its technological and strategic responses. However, in today's world—where not all risks can be anticipated or prevented—experiences with meteorological disasters serve as important examples both for those who produce such risks and for the disadvantaged states and communities that are most severely affected by environmental injustice, in their shared effort to build a safer world.

REFERENCES

AFAD. (2018). *Türkiye'de afet yönetimi ve doğa kaynaklı afet istatistikleri.* https://www.afad.gov.tr/kurumlar/afad.gov.tr/35429/xfiles/turkiye_de_afetler.pdf

AFAD. (2021). *Batı Karadeniz bölgesinde yaşanan sel afeti sonrasında bölgedeki çalışmalar devam ediyor.* https://www.afad.gov.tr/bati-karadeniz-bolgesinde-yasanan-sel-afeti-sonrasinda-bolgedeki-calismalar-devam-ediyor

AFAD. (2022). *AFAD hakkında.* https://www.afad.gov.tr/afad-hakkinda

Akbal, A. (2023). Doğal afetler ve afet yönetiminin sosyolojik değerlendirmesi. In M. A. Kasapoğlu (Ed.), *Afet ve toplum: İnsan doğa ilişkisi* (pp. 10–23). Töz.

ASHB. (2021a). *Bakanımız Yanık: "Başta Kastamonu olmak üzere Batı Karadeniz'i etkileyen sel felaketi 3. seviye afet ilan edilmiştir".* https://www.aile.gov.tr/haberler/bakanimiz-yanik-basta-kastamonu-olmak-uzere-bati-karadenizi-etkileyen-sel-felaketi-ucuncu-seviye-afet-ilan-edilmistir/

ASHB. (2021b). *2021 E-bülten: Eylül–Ekim 2021.* https://www.aile.gov.tr/media/96243/bi-ri-mbultenleri-baski-1.pdf

ASHB. (2022). *2021 faaliyet raporu.* https://www.aile.gov.tr/media/100242/2021-yili-faaliyet-raporu.pdf

Beck, U. (1992). *Risk society: Towards a new modernity* (M. Ritter, Trans.). *Sage (Atlanta, Ga.).*

Beck, U. (2010a). Climate for change, or how to create a green modernity? *Theory, Culture & Society, 27*(2–3), 254–266. DOI: 10.1177/0263276409358729

Beck, U. (2010b). How climate change might save the world. *Development and Society, 39*(2), 169–183.

Bilgen, G., Balcı, E., & Kalça, M. Y. (2022). Kastamonu Bozkurt ilçesinde 11.08.2021 tarihinde meydana gelen sel felaketinin yerinde incelenmesi. *Tasarım Mimarlık ve Mühendislik Dergisi*, *2*(1), 20–35. https://izlik.org/JA36EX57DJ

Bozkurt Municipality. (2010). *Bozkurt: Güzellik onun doğasında.* Asır.

Bullard, R., & Wright, B. H. (1987). Environmentalism and the politics of equity: Emergent trends in the Black community. *Social Thought & Research*, *12*(2), 21–38. DOI: 10.17161/STR.1808.5017

Bullard, R. D. (2000). *Dumping in Dixie: Race, class, and environmental quality* (3rd ed.). Westview Press.

Bullard, R. D. (2001). Environmental justice in the 21st century: Race still matters. *Phylon*, *49*(3/4), 151–171. DOI: 10.2307/3132626

Cordner, A. (2021). Risk. In Caniglia, B. S. (Eds.), *Handbook of environmental sociology* (pp. 167–187). Springer., DOI: 10.1007/978-3-030-77712-8_9

Japan, C. W. S. (2019). Towards mabi's recovery: lessons on year on. Retrieved from: https://www.cwsjapan.org/wp-content/uploads/2019/09/TowardsMabisRecovery_S.pdf

Japan, C. W. S. (2023). Lessons from Mabi: Five years of recovery form the 2018 Western Japan Flood. https://www.preventionweb.net/publication/lessons-mabi-five-years-recovery-2018-western-japan-flood

Dauvergne, P. (2009). *Historical dictionary of environmentalism.* Scarecrow Press.

Demirbilek, M. (2016). Environment, environmental refugees and green social work. *International Journal of Social Sciences and Education Research*, *2*(3), 905–914. DOI: 10.24289/ijsser.279033

Des Marais, E. A., Bhadra, S., & Dyer, A. R. (2012). In the wake of Japan's triple disaster: Rebuilding capacity through international collaboration. *Advances in Social Work*, *13*(2), 340–357. DOI: 10.18060/1964

Directorate of Climate Change (2024). *Paris Anlaşması.* https://iklim.gov.tr/paris-anlasmasi-i-34

Disaster Management in Japan. (2022). *Consecutive disasters: Toward the establishment of a disaster-conscious society*. https://www.bousai.go.jp/en/documentation/white_paper/pdf/SF1-1.pdf

Dominelli, L. (2012). *Green social work: From environmental crises to environmental justice*. Polity Press.

Dominelli, L. (2014). Promoting environmental justice through green social work practice: A key challenge for practitioners and educators. *International Social Work, 57*(4), 338–345. DOI: 10.1177/0020872814524968

Dominelli, L. (2015). Greening social work. In *International encyclopedia of social and behavioral sciences* (2nd ed.). DOI: 10.1016/B978-0-08-097086-8.28037-9

Dominelli, L. (2024). *Social work practice during times of disaster: A transformative green social work model*. Routledge.

EM-DAT. (2023). *2022 disasters in numbers*. https://www.cred.be/sites/default/files/2022_EMDAT_report.pdf

EM-DAT. (2024a). *Disaster year in 2023* (Issue No. 74). https://files.emdat.be/2024/04/CredCrunch74.pdf

EM-DAT. (2024b). *EM-DAT overview*. https://www.emdat.be

Encyclopedia of Philosophy. (2024). *The American environmental justice movement*. https://iep.utm.edu/enviro-j/

Erickson, C. L. (2018). *Environmental justice as social work practice*. Oxford University Press.

Euronews (2021). *Sosyal adalet endeksi*. https://tr.euronews.com/2021/11/17/sosyal-adalet-endeksi-turkiye-ab-ve-oecd-ulkeleri-aras-nda-sondan-2-s-rada

Görgün Baran, A. (2023). Depremin medyadaki yansımaları. In M. A. Kasapoğlu (Ed.), *Afet ve toplum: İnsan doğa ilişkisi* (pp. 83–92). Töz.

IASC. (2007). *Guidelines on mental health and psychosocial support in emergency settings*. https://interagencystandingcommittee.org/sites/default/files/migrated/2020-11/IASC Guidelines on Mental Health and Psychosocial Support in Emergency Settings %28English%29.pdf

IFSW (2020). *Global standards for social work education and training.* https://www.ifsw.org/global-standards-for-social-work-education-and-training/

INFORM (2025). *Results and data.* https://drmkc-jrc-ec-europa-eu.translate.goog/inform-index/INFORM-Risk/Results-and-data

IPCC (2007). Climate Change 2007: *Synthesis report.* https://www.ipcc.ch/site/assets/uploads/2018/02/ar4_syr_full_report.pdf

Japanese Red Cross Society. (2018). *Response to Japan floods and landslides 2018.* https://www.jrc.or.jp/english/relief/180724_005361.html

Karagüzel, , R., Türkkan Karaoğlu, G., Yılmaz, K., K. Çatlıoğlu, B., Çan, T & Akay, S., S. (2021). Kastamonu-Bozkurt 11 Ağustos 2021 taşkını ve düşündürdükleri. *Yer Mühendisliği*, *16*(October), 14–25.

Karataş, K. (2023). *Afet yaşamış toplumlarla çalışma* [Video]. YouTube. https://www.youtube.com/watch?v=cNQSHhPnldc

Lin, S.-S., Zhang, N., Xu, Y.-S., & Hino, T. (2020). Lessons learned from catastrophic floods in western Japan in 2018: Sustainable perspective analysis. *Water (Basel)*, *12*(9), 1-15. DOI: 10.3390/w12092489

LiveScience. (2011). *Japan's explosive geology explained.* https://www.livescience.com/30226-japan-tectonics-explosive-geology-ring-of-fire-110314.html

Maung, R., & Pellow, D. N. (2021). Environmental justice. In Caniglia, B. S. (Eds.), *Handbook of environmental sociology* (pp. 35–52). Springer., DOI: 10.1007/978-3-030-77712-8_3

McFarlane, A., C., & Norris, F. H. (2006). Definitions and concepts in disaster research. In Norris, F. H. (Eds.), *Methods for disaster mental health research* (pp. 3–19). Guilford Press.

Miranda, M. L., Hastings, D. A., Aldy, J. E., & Schlesinger, W. H. (2011). The environmental justice dimensions of climate change. *Environmental Justice*, *4*(1), 17–25. DOI: 10.1089/env.2009.0046

Miyamori, D., Yoshida, S., Kashima, S., Koike, S., Ishii, S., Okazaki, Y., Ikeda, K & Matsumoto, M. (2023). How the 2018 Japan floods impacted nursing home admissions for older persons: A longitudinal study using the long-term care insurance comprehensive database. *Journal of the American Medical Directors Association, 24(3)*, 368–375e1. DOI: 10.1016/j.jamda.2022.11.021 PMID: 36587929

Olağandışı Durumlarda Sağlık Hizmetleri. (2021). *Kastamonu–Sinop sel afeti 1. ay raporu*. https://www.ttb.org.tr

Oxfam (2011). *Gender mainstreaming: Putting women's rights at the heart of all we do*. https://oxfamilibrary.openrepository.com

Pak, M. (2023). *Afetlerde psikososyal destek hizmetlerinin etkisi: Sinop Ayancık örneği* (Unpublished doctoral dissertation). Ankara University.

Public Health England. (2020). https://assets.publishing.service.gov.uk/media/61f933b1d3bf7f78e469cd77/phe-annual-report-and-accounts-2020-to-2021-web-accessible.pdf

Soğancı, E. (2024). *Selden etkilenen topluluklara yönelik psikososyal destek hizmetleri: Bozkurt örneği* (Unpublished doctoral dissertation). Ankara University.

Stevenson, N. (2017). Ulrich Beck. In Turner, B. S., & Elliott, A. (Eds.), *Çağdaş toplum kuramından portreler* (pp. 435–449). İletişim.

Thomas, R., & Healy, L. (2010). Be prepared. In Gillespie, D., & Danso, K. (Eds.), *Disaster concepts and issues* (pp. 255–270). CSWE Press.

Tsosie, R. (2007). Indigenous people and environmental justice. *University of Colorado Law Review*, *78*(4), 1625–1678.

Türk Kızılayı. (2022). *Kızılay Bozkurtlu ailelere desteğini sürdürüyor*. https://www.kizilay.org.tr/Haber/HaberArsiviDetay/6593

Turner, B. S. (2006). *Sociology*. Cambridge University Press.

UNDRR. (2023). 2022 Global Natural Disaster Asessment Report. https://www.preventionweb.net/publication/2022-global-natural-disaster-assessment-report

United Nations (2015). *General Assembly resolution adopted on 3 June 2015.* https://www.preventionweb.net/files/resolutions/N1516716.pdf?startDownload=true

EPA (2024). *Learn about environmental justice.* https://www.epa.gov/environmentaljustice/learn-about-environmental-justice

Vilá, O., Cutts, B., Whitney, K., & Rivers, L. (2023). Environmental justice in disaster recovery: Recognition of the Latinx community by nonprofit leaders. *Climate Risk Management, 40.* DOI: 10.1016/j.crm.2023.100502

World Bank. (2019). *Adaptive social protection and disaster risk management: A case study in Japan.* https://openknowledge.worldbank.org

World Economic Forum. (2018). *This is why Japan's floods have been so deadly.* https://www.weforum.org/stories/2018/07/japan-hit-by-worst-weather-disaster-in-decades-why-did-so-many-die/

Yalçın, G. (2020). *Toplumsal cinsiyet perspektifinden doğal afetler* (Unpublished master's thesis). Hacettepe University.

Yeşilyurt, M. (2023). *Jandarmanın doğal afetlerdeki önemi: Bozkurt sel felaketi örneği* (Unpublished master's thesis). Kastamonu University.

Yoshida, S., Kashima, S., Okazaki, Y., & Matsumoto, M. (2023). Effects of 2018 Japan floods on healthcare costs and service utilization in Japan: A retrospective cohort study. *BMC Public Health*, *23*(1), 1-10. DOI: 10.1186/s12889-023-15205-w PMID: 36755264

Chapter 8

The Role of Microgrids in Addressing Climate Challenges and Revolutionizing Energy Systems

Jyoti Rani
https://orcid.org/0000-0002-5577-7808
Parul University, India

Ramratan Guru
https://orcid.org/0000-0001-6142-2643
Mody University of Science and Technology, India

Sakthivel Santhanam
Presidency University, India

Dharmjeet Prasad
Lovely Professional University, India

Shubhang Jha
https://orcid.org/0009-0008-1741-8389
Bennett University, India

DOI: 10.4018/979-8-3373-9685-9.ch008

ABSTRACT

Access to affordable, reliable, and clean electricity is vital for economic growth, social well-being, and sustainable development. However, conventional power systems face growing challenges, with climate change posing significant risks due to their high carbon footprint. Frequent blackouts, reduced efficiency, and infrastructure damage highlight the urgent need for solutions. Power utilities must modernize infrastructure and adopt low-carbon energy sources. This chapter explores how microgrids can address these challenges by offering decentralized, sustainable power generation that enhances resilience and reliability. Microgrids support climate adaptation and mitigation efforts, ensuring energy access despite environmental uncertainties. The chapter highlights their benefits, including improved efficiency and reduced emissions, while also addressing technical, economic, and regulatory challenges that hinder widespread adoption. By leveraging microgrids, communities and industries can transition to a resilient, sustainable energy future.

INTRODUCTION

Electricity unlocks sustainable development that can deliver economic growth, social inclusion and environmental sustainability. Access to electricity is a key driver of many socio-economic indices (i.e., education, healthcare, job creation) and is critical to facilitate improvements in the quality of life and poverty alleviation. During this article, we are focusing on the limitations of renewable energy integration, financial development and technology innovation to enhance electricity supply at large, mainly in developing and emerging economies. Consequently, the following sections, use the information presented in each of the given papers to explain the role of electricity in the sustainable development process.

Economic Growth & Social Inclusion

This new access to electric light during the night is very important for many different types of economic enterprises, particularly economic enterprises in rural areas of the country, where investments in grid-delivered electricity are often not viable due to low demand resulting from poverty levels. These Models of smart grid designs which insinuate socioeconomic status of rural consumer at its core will participate in triggering economic sustainability and social inclusion (Onu *et al.*, 2024).

In low- and middle-income nations, every 1% of GDP and financial development translate into a 0.15% and 0.14% boost in electricity access respectively. In this context, the positive effect of economic growth on its gradual enhancement of electricity is notable (Li, 2024).

Clean Energy and Technological Advancements

Providing access to electricity generation and clean fuels can expand electricity access, highlighting the role of clean energy in sustainable development (Li, 2024).

As long as ethical and regulatory issues are tackled, AI and digitization in the energy sector will help to accelerate universal access to clean and affordable energy (Awogbemi, Kallon and Kumar, 2024).

Reliability and Efficiency

Reliable electricity is critical for economic development, and poverty reduction. Resource Constraints Instead of Utilization Blocks Universal access to reliable electricity in East Africa (Nsabimana *et al.*, 2024).

With solar now accounting for around one tenth of energy consumption, its reliability is ever more pressing in relation to energy poverty - which speaks to Citroën's intention to expand to small van style references for systems that guarantee reliability for the solar home systems of national grids that are more common in larger vehicles. The effectiveness of their household electricity has a great impact on electricity affordability (Ferrall, Callaway and Kammen, 2022).

Connect is often highlighted, but electricity must also be reliable and affordable. Addressing these obstacles will be critical to ensuring electricity access can be a powerful force for sustainable development goals, combining policy measures, technology innovation and financial investment.

Challenges Faced by Conventional Power Systems, With a Focus on Climate Change Impacts

Climate change affects conventional non-renewable power systems in several ways; from demand for electricity generation to the supply side. So, the challenges are, the economics of planning and operations have to change a lot to give reliability and efficiency.' The impacts of climate change on power systems are higher temperatures, changes in precipitation and intensity of weather events causing capacity deratings, increased demand and vulnerabilities in infrastructure. The following sections provide more detail about these impacts and adaptations.

Demand-Side Impacts

Higher Electricity Consumption: As it gets hotter climate change is probably going to directly translate into more electricity usage, especially on summer months when more cooling will be required. This is also true of the Southeast U. S. where installed capacity is projected to increase by 35% 2050 in order to meet rising demand at a cost of substantial investment (Ralston Fonseca *et al.*, 2021).

Demand Shock Potential: The central U.S. shows that gaps of 5% to 21% of supply could occur, depending on climate severity, through climate change-induced demand increases. This underscores the importance of including climate projections in energy plans, to avoid potential supply shortages (Gurgel *et al.*, 2024).

Supply-Side Impacts

Derating of Thermal Capacity - Limitation on operational capacities of thermal power plants when sourced water for thermal power plants is heated and also the atmospheric temperatures increase It may cause loss of efficiency and capacity and lead to the necessity for design alterations such as improved cooling mechanisms to maintain operation (Loew *et al.*, 2020).

Renewable Energy Variability: Climate change can also influence when and how much renewable energy resources such as hydro, wind and solar can produce. For instance, hydroelectric power is sensitive to fluctuations in precipitation and water availability, and the outputs of wind and solar are also contingent on weather patterns (Zhang *et al.*, 2022).

Infrastructure Vulnerabilities

Extreme Weather Events: Events such as storms and heatwaves cause damage to the existing infrastructure which affects the power supply especially the distribution networks. These incidents lead to large-scale outages and loss of revenue to utilities in Indonesia.

Water Resource Constraints: Because a significant percentage of total U.S. electricity generation comes from water-cooled thermal power plants, the quantity of available water and the temperature of supplied water can influence generation of electricity across the nation. Therefore, regional adaptation measures are necessary in order to ensure energy security and system reliability.

Adaptation Strategies

Invest in renewables Increasing renewables in the power mix helps to fill supply gaps and reduce carbon emissions. Larger shares of renewable sources reduce the potential supply gaps in the central US (Gurgel *et al.*, 2024).

Integrated planning frameworks: Constraints posed to the water-energy nexus for climate While this solution requires greater investment in the capacity and infrastructure required, is a long-term solution that fosters sustainability.

Resilience Improvements: Development of integrated resilience frameworks that includes social and technical aspects can help the power system better respond to large scale climate events.

If the challenges of traditional power systems presented by climate change are daunting, they also present pathways for innovation and adaptation. Integrating climate into planning and operations can make power systems more resilient and sustainable. But this will require huge investments and a shift to more flexible and adaptable energy systems. However, the transition to a low carbon energy system complex as it may be is essential to mitigating the effects of climate change and to powering a sustainable future

Overview of the Role of Microgrids in Addressing Energy Sector Challenges

Microgrids demonstrate how a universal answer, even when quite an overall reach past the current ground zero, delivers to a world facing a dynamic, transformative moment in the electric energy industry in just about every corner of the planet as more overwhelming uptake of the molecular model converges to a new frontier of grid integration, sustainability. This renders them a promising solution to couple distributed energy resources and demand-side management for electricity grids, while standing among the required pillars for a sustainable energy transition (Lopez Alzate, Gómez-Luna and Vasquez, 2024). This decentralization of power generation contributes to local resilience and sustainability in the face of climate change and environmental challenges (Farhan Mukhtiar, 2024) Microgrids, when well-managed, increase the efficient utilization of renewables (wind or solar) and energy storage system, helping to reduce costs and carbon emissions. Additionally, advanced technologies like AI, ML and real-time data analytics can enhance microgrids' speed and reliability (Farhan Mukhtiar, 2024) (Irmak, Kabalci and Kabalci, 2023). Additionally, microgrids enhance and benefit grid stability and load balancing, reducing the problems caused by the increasing integration of renewable energy and electric vehicles. Multi-microgrid-based systems and blockchain-based energy sharing/trading mechanisms significantly improve the operational and economic efficiency of smart distribution grids (Kumar *et al.*, 2021) (Lei *et al.*, 2024). Also, the digitalization of the microgrids by means of Internet of Things(IoT), Big Data and big data analytics and Cybersecurity systems are too important for fulfilling and monitoring the increased energy demand, while ensuring the system reliability and security (Irmak, Kabalci and Kabalci, 2023). Modern grids are a source of energy efficiency and economic savings; alongside the increasing rate of sustainable solutions a more modern approach becomes imperative (Zheng *et al.*, 2024) (Ahmed *et al.*, 2023).

CLIMATE CHANGE AND THE POWER SECTOR

The Carbon Footprint of Conventional Energy Systems

Usage of traditional energy systems in particular improved systems based on fossil fuels releases greenhouse gases, which in turn are the main driving factor for climate change. With the power sector, at least four-fifths of which is based on fossil fuels, responsible for nearly 40% of global energy-related CO2 emissions every year. From the need to tackle the effects of climate change, this much; it will require us to shift to low-carbon technologies.""" So those findings, which essentially prove the carbon footprint of conventional energy systems and low-carbon alternatives, are in different areas everyone knows that we are facing inflection points, complexities and challenges in managing this transition. Below, the carbon footprint of conventional energy systems and how the power sector is relevant to the phenomenon of global warming.

Conventional Energy Systems Carbon Footprint

Depending on Fossil Fuels: The world still depends on fossil fuels since the Industrial Revolution and the carbon emissions have increased. The emissions range of fossil fuel power plant carbon footprints is very broad, it was reported, for example, that combined cycles plants using natural gas emit 321–980 g CO2 eq/kWh, based on the plant type and fuel transmission (Dalir, Shafiepour Motlagh and Ashrafi, 2017).

Carbon Lock-In: The most worrisome issue is carbon lock-in that is, the economic, technical and institutional reasons why carbon-intensive technologies continue to exist. Such Lock-in effect greatly limits the possibility of switching to a low-carbon alternative, especially in sectors like coal and gas-based power plants.

Shift to Low-Emissions Technology

Renewables and CCS: The role of renewables, particularly wind and solar, and carbon capture and storage (CCS) are big ones, with cuts in carbon emissions. They are expected to be responsible for more than 50 percent of cumulative reductions in emissions by 2050 (Lee and Lin, 2019). However, the upfront infrastructure costs of renewables are still above those of fossil systems (Hertwich, 2021).

Energy Storage and Management: Lithium-ion batteries in energy storage systems decouple power generation from demand in a manner conducive to decarbonization. Localized energy systems will become less carbon-intensive owing to the implementation of efficient energy management protocols (Parlikar *et al.*, 2024).

Challenges and Considerations

Life-Cycle Emissions: With the transition to low-carbon technologies from conventional technologies, there is an increase in the initial emissions, followed by a visible decline in emissions over the operational lifetime. Evaluating such trade-offs, and the mitigatory potential of low-carbon technologies more generally, depends on life-cycle assessments (LCA).

Consumption-based carbon accounting: Accurate carbon accounting is the prerequisite for consumption-based low-carbon strategies to be implemented. This approach captures emissions temporal and spatial granularities, as well as the power grid operation constraints (Chen, 2024).

This transition to low-carbon energy systems is required to mitigate climate change but is not without its challenges. File the biggest barriers under carbon lock-in (the continuing use of carbon intensive technologies), oil- and coal- intensive upfront investment that limits renewables infrastructure, and the challenge to conveniently measure and keep records of carbon emissions. However, the ability of the power sector to deliver a sustainable energy future lies in the improvement of energy storage technology, carbon capture technologies and renewable technologies which are moving at a exponentially growing speed.

Impacts of Climate Change: Blackouts, Reduced Efficiency, and Infrastructure Damage

Climate change is posing a threat to infrastructure, which is a concern for blackouts, low performance and asset damage. The impacts are driven by the increasing frequency and intensity of severe climatic events, increasing temperatures of the air and water, and changing precipitation regimes, all impacting the reliability and performance of critical infrastructure systems. The following sections examine these impacts in detail with references to findings from multiple studies.

Blackouts

Increased frequency of blackouts: Climate change will make extreme weather like hurricanes, heatwaves, and storms more frequent and severe causing power outages. The number of major electrical grid failures has more than doubled over the past few years in the United States, with blackouts concurrent with heatwaves posing a considerable public health threat due to non-functioning air-conditioning systems.

The Risk of a Power Down: The electric grid is particularly vulnerable to extreme weather: When Superstorm Sandy hit New Jersey, the electric grid suffered for hundreds of miles, leading to blackouts that lasted for days. Future climate scenarios predict the occurrence of more extreme storms which would further add to the demand for power systems to be resilient.

Reduced Efficiency

Effect On Electric Transmission: Increasing air temperatures reduce the ampacity of electric transmission lines, limiting their ability to transmit electricity. Coupled with increased demand for electricity during warmer stretches, this can put strain on power infrastructure. In the U.S. transmission capacity could decline by 1.9%–5.8% by mid-century(2) and peak electricity loads may increase by 4.2%–15% (Bartos *et al.*, 2016).

Energy System Impacts — Climate change is more likely than not to increase cooling demand and decrease heating demand globally, and these changes are likely to affect the efficiency of energy systems. Regional variation does exist, but the most pronounced effects are predicted in South Asia and Latin America.

Infrastructure Damage

Flooding and coastal damage: Flooding, a major risk to infrastructure, has been projected to increase for energy users. Even with a global temperature increase limited to 2°C, the number of users reliant on electricity infrastructure vulnerable to flooding may double. A 4°C rise would treble flood damage in the UK (Dawson *et al.*, 2018).

Regional Variability in Europe: Danube and el Niño river data suggests damages due to climate extremes could be 10 times higher by the end of the century, with southerly and sou' easterly latitudes disproportionately affected. Analysis of the data show the economic losses predicted to be top-heavy towards the sector industries of the industry, transport, and energy(Forzieri *et al.*, 2018).

This Will Destroy U.S. Infrastructure: Roads, bridges and coastal developments in the U.S. face rising temperatures, shifts in precipitation and sea-level rise. These shifts may have substantial economic impacts, although adaptation initiatives could mitigate some of the impacts (Neumann *et al.*, 2015).

Climate change can make its impacts felt down to the communities from the global level, and there are opportunities for adaptation and mitigation. The effects on infrastructure sectors can be heavily diminished if cost-effective adaptation measures are implemented and aggressive mitigation policies are adopted (Neumann *et al.*, 2015). Finally, metrics for assessing infrastructure performance in the face of climate risk, including making assessment information available to stakeholders can assist practitioners to prioritize adaptation. However, the analysis provides a wealth of examples of the types of thorough risk assessments and coordinated adaptation strategies that these complex and interdependent infrastructure systems will require to address these problems head on whatever the new normal may be.

Urgency for Transitioning to Resilient and Sustainable Energy Systems

Urgent attention has been drawn towards the need of adopting resilient and sustainable energy systems due to the multidimensional threat of direct effects of climate change, economic collapse, and international instability. This transition is not just an environmental imperative, as we all know the planet is getting hotter, but a geopolitical transition that must happen if we are to secure our energy and economic future. IGUAS Integration of Renewables, Hacking Resilience and Game Changing Tech In the coming sections we summarize the major areas of this transition as per the research papers listed.

Digitalization and Resilience

Digitalization can help, too. Especially since we recently witness this happen to the COVID-19 pandemic, showing flaws in this type of grid system as well (Zakeri, 2024).

Innovation through big data analytics and digital technology: Renewable energy sources present challenges for grid integration, energy distribution and system reliability that data-driven models and digital tools can help address (Zakeri, 2024).

Integrated and Sustainable Energy Plans

Transdisciplinary approaches and frameworks for energy system planning will be necessary to address complex problems such as the ones of climate change and economic shocks. Whereby socio-economic aspects, environmental system design and stakeholder engagement are integrated (Koundouri *et al.*, 2024).

An instance of this manner uniformity is the global local climate hub initiative, which nurtures transdisciplinary approaches to create lasting methods of energy generation, and bring up the need for inventive sustainable renewable energy portfolis (Koundouri *et al.*, 2024).

Trade and Economic Sustainability and Resilience

RES has become a significant determinant for the sustainable use of the economy. RES reduces energy shocks, protects economic productivity and generates real savings (Fan, Wang and Lv, 2023).

Pathways for RES response are improvements to energy transition, infrastructure and increased research and development (Fan, Wang and Lv, 2023).

Environment and Energy Transition Policy

Environmental policies and energy transitions help reduce climate change impacts. The case of OECD countries suggest that strong policy interventions and investment in renewables can significantly lower CO2 emissions (Ma *et al.*, 2023).

The transition towards the broad spectrum of renewable energy should be implemented at the macro as well as micro level in order to reduce reliance on fossil fuels (Ma *et al.*, 2023).

Green Microgrids and Security of Energy

Green microgrids represent an eco-friendly option compared to conventional power generation systems, reducing dependency on foreign energy assets while also removing risks associated with political clashes (Kiehbadroudinezhad *et al.*, 2023).

Microgrid technology has developed to a point of being deployable in many rural areas, although there are still concerns about high costs and reliability (Kiehbadroudinezhad *et al.*, 2023).

Integrated Frameworks for Energy Transition

Cleantech renewable energy communities are decentralized energy systems that promote resiliency and decarbonization.

Energy transition planning is a multidisciplinary framework that addresses environmental, technical, social, economic, and political aspects of sustainable development.

Resilience-Driven Restoration

Such renewable generation can create an active distribution grid, thus new tools are needed to facilitate a resilience driven restoration.

Enhanced situational awareness tools enable rapid restoration of outages, in particular, to maintain some critical loads.

Energy Transitions Sparked by Catastrophe

Now disasters can also be upending catalysts to accelerate the transition to low-carbon, resilience-oriented energy systems. Innovations and rapid decarbonisation are more likely when they are set within post-disaster scenarios.

Local and regional responses to disasters can be integrated with national energy policy to further socio-spatial resilience.

But this is not without its challenges, even though it is crucial to the unavoidable movement towards resilient, sustainable energy systems. The difficulty is in the expensive nature of the development of resilient systems, making a need for policy frameworks as well as coordinating multiple energies. Moreover, this transition must be carefully managed to minimize negative unintended consequences such as energy sprawl, while also ensuring that benefits are equitably distributed across regions and communities.

UNDERSTANDING MICROGRIDS

Microgrids are localized energy systems that can operate independently or in conjunction with area grid power including generation sources like combined heat and power, wind and solar power, energy storage, and optimized demand response. MIt can train data on it until October 2023 Microgrid architecture consists of various renowned components being deployed under specific topologies and controls to achieve the aforementioned objectives.

Distributed Energy Resources (DERs): Utilities can also learn about smaller scale electricity generation technologies such as solar photovoltaic (PV) systems, wind turbines, micro-turbines, fuel cells, and diesel power plants. Distributed Energy Resources (DERs) are the heart of microgrid philosophy, which provide the essential energy source in the grid system (Abbasi *et al.*, 2023). Energy Storage Systems: Batteries, flywheels and other forms of storage technology store excess energy generated by DERs and provide power during periods of low generation or high demand. As in microgrid, energy storage is essential for balancing supply and demand (Abbasi *et al.*, 2023). Local Loads: Local loads key to understanding microgrid operations are the demands tied to the microgrid, typically of a residential, commercial or industrial nature; The microgrid can meet the requirements from locally generated energy or from the distributor as required (Jirdehi *et al.*, 2020). Control Systems: Advanced control systems are involved in the management and operation of the microgrid, ensuring stability and efficient performance. These systems are responsible for managing the operation of DERs, storage and loads, as well as for facilitating transitions between grid-connected and islanded mode (Abbasi *et al.*, 2023) (Ahmed *et al.*, 2023).

Microgrid Configurations and Topologies

AC & DC Configurations: A microgrid can be delivered with alternating current or direct current distribution buses. While AC systems might be more compatible with existing infrastructure, DC systems may have a more efficient potential for some use cases (Cabana-Jiménez, Candelo-Becerra and Sousa Santos, 2022).

Hybrid Configurations: Microgrids with both AC and DC systems which are coupled using interlinking converters which facilitate power sharing between them. The road map will contribute to increase flexibility of microgrid and enhance performance.

Topologies: Based on the topologies isolated, networked and multiple microgrids topologies. Similarly, this topology selection also affects microgrid reliability, efficiency, and cost (Cabana-Jiménez, Candelo-Becerra and Sousa Santos, 2022).

Benefits and Challenges

Benefits: These systems can offer many advantages, including better power quality, reduced losses on transmission lines, more integration with renewable energy sources, and enhanced resilience and security against grid disturbances (Abbasi *et al.*, 2023) .

Challenges: Despite their advantages, microgrids hold challenges such as high initial installation costs, complex control needs, and the need of its effective role on existing grid infrastructure (Ahmed *et al.*, 2023).

Microgrids hold promise for the energy landscape of the future, but they need proper planning and management if they are to reach that potential. And to improve performance of such modern systems, considerable improvement in renewable energy supply and control systems of such systems are needed to build a stable and economical grid. Microgrids will become more prominent as research and technology advance.

Key Characteristics: Decentralization, Sustainability, and Resilience

Therefore, Microgrids are recognised as the future of energy systems with respect to decentralised, sustainable and resilient energy systems. The solution comes with power operations relying on renewable energy resources in conjunction with smart control systems to remain sustainable and reliable. The features of microgrids, specifically decentralization, sustainability and resilience are crucial to addressing the energy dilemmas of the new era, ranging from climate change to energy security. The response below explores what those traits are, drawing on insights into some new research.

Decentralization

Distributed Energy Resources (DERs): Microgrids use DERs (e.g. solar panels and wind turbines) which transformed traditional centralized grids into distributed networks. This decentralized approach allows for localized energy generation and consumption, reducing dependence on centralized power plants and enhancing energy security (Kiehbadroudinezhad *et al.*, 2023).

P2P Energy: The sharing of excess energy provides a mechanism to increase the efficiency and flexibility of the system in P2P energy sharing in microgrids. This is a powerful means not only to optimize resources consumed, but it would mean less reliance on the main grid, which augurs well for a more decentralized energy structure.

Privacy and Competition: Because of the privacy concerns and the need to keep competition pressure in the market, DC-MGs are generally reluctant to share internal information about the configuration and operation of the microgrid with the distribution system operators. This is useful in emergencies where decentralized management allows for better resilience (Mansouri *et al.*, 2022).

Sustainability

Integration of Renewable Energy: Microgrids enable the integration of renewable energy sources like solar and wind, crucial for sustainable energy generation. This integration helps to advance the transition to carbon-neutral energy systems, as part of the global sustainability targets (Farhan Mukhtiar, 2024) (Kiehbadroudinezhad *et al.*, 2023).

Energy Storage Systems: For microgrids, the use of distributed energy storage systems is a key factor for adjusting power supply and demand, which in return enhances sustainability by enabling renewable energy usage.

Cost and Technological Developments: The cost-effectiveness development of innovative constructions and clean microgrid technologies underlies the implementation of sustainability energy systems that would be especially required in remote and rural communities (Kiehbadroudinezhad *et al.*, 2023).

Resilience

Resilience Metrics and Optimization for Microgrids: Following different approaches, Microgrids use the distributed optimization algorithms for resilience-related optimization as well as the resilience metrics. Such approaches help reduce the uncertainties of systems and enhance the fault tolerance capability of the systems against the occurrence of faults or disturbances.

Resilience-centric Collaboration: The networked microgrids (NMG) ADD to Dictionary Architectures empowered by decentralized structures and multi-agent reinforcement learning to improve resilience. A cost-effective coordination among microgrids can be achieved through a distributed framework which will improve resilience of microgrids horizontally against extreme events.

Demand Response and Digital Twin Technology: the demand response program and digital twin technologies can be implemented in the microgrid to enhance microgrid resilience with advanced simulation of worst-case scenarios, optimizing operations in the real-world (Cao and Zhou, 2024).

Although decentralization, sustainability, and resilience boost microgrids, they indeed come with challenges. These include significant capital investments, tech complexities, and a need for proper control mechanisms to manage decentralized resources efficiently. Implementing microgrid solutions in conjunction with existing energy systems will also require the development and implementation of appropriate regulatory frameworks and market dynamics. Despite these challenges, though, microgrids hold great potential for energy needs of the future, largely owing to its benefits.

Advantages Over Traditional Power Systems

Microgrids also offer relative advantages over traditional centralized power systems that can help address important challenges related to energy reliability, sustainability, and resilience. Their primary benefit is improved energy resilience. Microgrids, in contrast to centralized grids, can operate in an islanded mode to protect key facilities during a disruption, which could also be wide-spread outages from natural disasters or system failures. The integration of renewable energy sources such as biomass, wind, and solar are another significant benefit to district systems by reducing reliance on fossil fuels and ultimately removing greenhouse gas emissions. This makes microgrids a sustainable solution, which meets the worldwide commitment to mitigate climate change. Because power is produced and distributed in the same locality, microgrids reduce the losses related to energy transmission, improving efficiency and reducing costs. Microgrids extend power access to remote or underserved communities where extending the main grid is not cost-effective or feasible. This characterized the need to change to a more localized method of implementing sustainable development in communities, which helped communities to become self-sustained that helped propel their economic development as well. In addition, advanced technologies like smart controllers and energy storage devices enable microgrids to utilise energy and align consumption with generation to reduce waste and operational expenditure. The advantage of microgrids, due to their decentralized power generation capabilities, is that the full system remains reliable, less susceptible to cyberattacks and has better potential to develop and adapt with the energy landscape in flux, making this an exciting alternative to the bulk generation systems used now.

MICROGRIDS AND CLIMATE CHANGE

Role in Climate Adaptation: Enhancing Energy Resilience

This is particularly the case in the climate adaptation realm where microgrids serve a role in bolstering the energy resiliency. And as climate change makes force majeure events more frequent and severe, we need more robust energy systems. Microgrids, whose ability to operate separately from the central grid shows promise as an energy provider during and after these types of occurrences. They use a mix of green energy sources and advanced technology to deliver stable and reliable energy. This response explores the adaptation potential of microgrids based on their resilience-enhancing characteristics.

Leveraging Advanced Technology to Bolster Resilience

Combining Machine Learning and Optimization: The usage of ML algorithms (e.g. the Hybrid Autoregressive Integrated Moving Average-Deep Reinforcement Learning (HARIMA-DRL)) will greatly boost the resilience of the microgrid. It also enhances the grid's operational efficiency and its ability to restore power following disasters, resulting in resilience values that outperform other conventional models, like CNNLSTM and Gauss Reinforcement Memory GRM) (Kumar and Kumar, 2024).

One of these technologies that can facilitate the microgrid objectives is Digital Twin technology. The technology sends a living virtual twin of the grid that can apply mitigation strategies and proactively strengthen grid resilience (Cao and Zhou, 2024).

Incorporating Renewable Energy Sources

Mobile Wind Turbines and Hydrogen Storage: The presence of MWTs and HSSs at the microgrid provide spatiotemporal flexibility in power supply and thus improves resilience. This dead-cycle strategy helps to eliminate the kill threats of HILP and contributes to the decarbonization of energy systems (Su *et al.*, 2024).

Green Microgrids: Switching to green microgrids that rely on renewable energy sources reduces dependence from conventional power systems and enhances the ability to withstand uncertainties like political wars and pandemics. These systems provide valuable benefits over sustainable development goals and the Paris agreement to produce clean energy (Kiehbadroudinezhad *et al.*, 2023).

Consideration of Socioeconomics and Urban Planning

Integrated Microgrid Planning: MicrogridsPLus is a remote monitoring, diagnostics and delivery capability for deploying and operating battery energy storage. Such a methodology addresses socioeconomics, preventing disasters and supporting aspects of sustainable urban change (Ottenburger *et al.*, 2024).

Dynamic Boundaries and Network Reconfiguration: Dynamic boundaries enable microgrids to change (expand) depending on the situation; this flexibility, allows them to enhance resilience whether supplying critical loads to withstand disturbances. Techniques like Mixed-Integer Linear Programming (MILP) and network reconfiguration facilitate this agility by optimizing microgrid operation.

Challenges and Future Perspectives

Estimate and manage uncertainties: Given that rooftop production and demand on the distribution network can come with uncertainties, robust optimization techniques must be available to demonstrate microgrid resiliency. Two techniques that facilitate the effective management of these uncertainties are hybrid stochastic-robust optimization (HSRO) and scenario-based methods (Cao and Zhou, 2024).

Multi-agent Reinforcement Learning: The decentralized approach utilizes MARL effectively for coordinating the NMGs as one coordinated entity, facilitating resilience in case of NMGs through appropriate credit assignment to each NMG for their contribution towards system resilience.

Such a method removes the drawbacks of centralized distribution management.

As a potential solution for the energy resilience challenges many organizations face, microgrids are not to be taken lightly, as they have had their own cost and underlying reliability challenges. Green microgrids are a solution to overcome these challenges to generate sustainable and clean power. Additionally, the development of technologies and integration of renewable energy sources must also factor in socioeconomic considerations to foster more equitable and resilient energy systems.

Case Studies Showcasing Successful Microgrid Implementations

Microgrids have recently emerged as a key enabling technology in this new data hosting paradigm, enhancing both energy management and grid integration in creep regions that lack grid connectivity. These successful microgrid deployments demonstrate that commercially viable, affordable, and resilient energy delivery is possible. In this answer we will dig into some examples of international microgrids showcasing successful installations in a range of locations and contexts.

Innovative Microgrid Services in Colombia

This article is part of a global case study series led in Colombia examining the role of microgrids in advancing grid resilience and reliability. Special emphasis is placed on ancillary services improvement by means of DERs and a spatially distributed control methodology. This also improves grid stability and enables a shift towards the use of renewable energy (Lopez Alzate, Gómez-Luna and Vasquez, 2024).

Data-Driven Microgrid Selection in Afghanistan

In northern Kandahar City, Afghanistan, one prototype of a mini-grid built to address the energy needs of a community has been operating. This energy system consists of solar PV, diesel generators, and batteries.

Using HOMER software however, to determine the optimal configuration we found net present cost of 10,108andcostofenergyat0. 75/ kWh, demonstrating the utility of microgrids even in low-income areas (Tayyab *et al.*, 2024).

A Single-Controllable Microgrid

A microgrid controlled by a single entity, consisting of commercial devices, was installed on the campus of the Federal University of Minas Gerais, Brazil. This microgrid fulfills the demand in islanding or grid-connected mode through self-consumption, time-shifting, or peak-shaving. Energy storage of the system includes lead-acid, lithium-ion and sodium-nickel batteries (lithium-ion being most efficient). This case study illustrates the benefits and the real-world limits of deploying microgrids (Reis *et al.*, 2022).

Microgrid Planning in Chile

Identifying planning uncertainty in renewable resources, the method of planning developed, was applied to a rural Mapuche community in Chile. A microgrid configuration premised on the new framework towards microgrid optimization in furtherance of computational intelligence techniques resulted in an investment risk increase of 19.9%. This one offers hands-on field experience that focuses on the pre-feasibility of microgrid projects in rural locations with implications to outcomes (Morales *et al.*, 2024).

Multi-Microgrid system in the Democratic Republic of the Congo

To enhance energy reliability and reduce costs, a multi-microgrid system was proposed and established in DRC's Goma. The system is comprised of solar PV, methane gas captured from Lake Kivu, battery storage on site, and the potential to push energy back into the grid when needed. Interconnected way ensures system stability and enlightens resources' sharing of microgrid (Khan Ngwashi *et al.*, 2023).

These are successful case studies of microgrid implementation; however, challenges still exist, such as ensuring careful planning and smart operation due to the intermittent nature of renewable energy sources. Furthermore, the adaptation of microgrids with the current energy framework poses a different kind of challenge that needs consideration of technological and economic elements for durability and reliability.

DEVELOPMENT AND DEPLOYMENT CHALLENGES

Data is available under Open Data Commons Attribution-ShareAlike License; additional terms may apply. The same legal and regulatory frameworks can promote or inhibit microgrid projects, acting as either enablers or obstacles through how well they help to address issues such as financial and technical barriers, legal ambiguities, and other matters. This has a number of impacts on the microgrid policy implementation and operation angle.

Economic Viability and Price Signals

Regulatory policies including feed-in tariffs (FiT), net metering (NM) and energy storage incentives (ESI) have a strong impact on the feasibility economics of microgrids. For instance, employing dynamic FiT frameworks has been shown to enhance microgrid reliability and decrease operational expenses by justifying the effective utilization of battery energy storage systems (BESS) (Gholami, Muyeen and Abokhamis Mousavi, 2024).

For example, income tax deductions, accelerated depreciation, VAT exemptions, customs duty exemptions, and the such can help in improving the financial viability of microgrid projects. Such incentives lead to positive cash flows and make microgrids financially attractive (Gomez-Garcia, Askar-Rodriguez and Velasco-Medina, 2021).

Legal and Regulatory Uncertainty

Microgrids operate under four different authoritative backdrops throughout Europe; without a harmonised allied regulatory framework, many uncertainties emerge that could hinder microgrid implementation. The situation becomes trembling when incorporating microgrids into the EU electricity market where national regulations vary significantly among countries and where the ability of the member states to grant energy communities operating rights on distribution networks (Behrendt, 2023).

Brazil is an early adopter of microendr markets due to its existing regulations allowing microgrids to connect to the distribution system, but the absence of clear and specific policies for microgrids has limited large-scale deployments. In conclusion, one of the motivations for the proposed regulatory changes is to create a fairer and more consistent regime that will benefit all parties involved in distributed generation.

Technical and Operational Limits

Microgrids have regulatory structures that may impose technical limits on their design and operation. For instance, it has also been shown to support performance and economic feasibility in the planning of microgrids through the incorporation of regulatory constraints within the energy-management systems of these microgrids .

It is also necessary for the microgrid designs to take into account regulatory constraints in terms of renewable energy sources, consumption habits and market prices. Stochastic optimization models that account for these factors can enhance the reliability and economic performance of microgrids (Alonso *et al.*, 2020).

Security and Resilience

Cyber security needs through primary policies that define the safe operation of the microgrids next to the smart grid, which could be more closely integrated. Operation-focused decision making frameworks can help mitigate cyber threats and improve the resilience of urban microgrids to such threats (Lu, Xu and Ma, 2024).

Regulatory frameworks and policies can act as enablers of microgrid development, but also as drivers of challenges. Existing issues of reliability and security in the electricity generation sector also limit the mainstream adoption of microgrids. In this regard, the technical, economic and security complexities of integrating microgrids into the already existing energy systems need to be considered. If a completely secular market can impact these burthens using complete, and adequate regu latory drives, a future labor-prosperous complex of microgrids can be reconciled to account for the vigor changes necessary for a sustainable future growth.

STRATEGIES FOR OVERCOMING CHALLENGES

Technological Advancements for Seamless Microgrid Integration

It is similar in power of microgrid integration in future energy systems, pioneering progress also continues on. They are connected to control strategy improvement, synchronisation, power management and accessory security, vital for microgrid performance and stability. The subsequent sections cover key innovations enabling such low friction microgrid integration.

Advanced Control Strategies

Introduction of Virtual Synchronous Generators (VSGs): The VSG-based inverters can increase its adaptive stable capacity of a power system by emulating synchronous generator control strategy. This approach allows for smooth transition between off-grid and on-grid operation and provides inrush current during synchronization between 120A and nominal values while improving frequency stability (Tahir *et al.*, 2024).

Petri Recurrent Wavelet Fuzzy Neural Network (PetriRWFNN): Controller has also been proposed in order to improve the transient responses of the active power and voltage in different operating modes of microgrids. It serves as a grid reconnection presynchronization estimation method and allows for seamless mode transition from islanded mode to grid-connected mode.

Solutions for Power Management and Storage

Hybrid Energy Storage Systems (HESS): The integrated wind and solar energy and interleaved HESS serves as the real time power controller for LVDC microgrids. A system will require such systems to demand the requirements of piezoelectric energy harvesting system and provide the required input power continuously from the battery widely produced by a system which can dynamically adjust its energy management system catering to the variable sources of energy and the incoming load within the system input through balancing the power of system thereby reducing stress on the battery enabling the newer technology applications such as energy autonomous systems and self-sustaining embedded devices (Behera and Pattnaik, 2024).

Synchronous Reluctance Generators (SynRG): Used in microgrids for smooth transition of power, sensor less field-oriented control with data-dependent compensation improves synchronization and dynamic response under weak grid conditions.

Synchronization and Connection to the Grid

SOGI-FLL-WDCRC Filter: This a synchronization technique for AC microgrids which contributes to power quality improvement, providing instantaneous and precise estimates of phase angles and frequencies despite grid abnormalities. It lowers harmonics and reactive power problems and produces high-quality power.

Master-Slave Control with Robust Algorithms: In AC microgrids, a master-slave architecture was developed for operating, at the customer side, the grid-connected and isolated states with robust control algorithms. Methods such as optimal support vector machine and Kalman filters improve islanding detection and grid synchronization.

Community, Identity, and Communication Enhancements

Microgrids use IoT protocols for bidirectional communication, but this poses security threats. The DDS protocols an RBAC-based security method increases cyber resilience via limiting access as well as minimizing communication attack vectors.

Experiments and Case Studies

Laboratory obtained to test the laboratory benches: Some of the studies verified the proposed technologies through laboratory benches, as they showed that synchronization, power management, and control in dynamic scenarios was being performed efficiently.Results of laboratory. The practical applicability and appropriateness of the proposed solutions are demonstrated empirically through these experimental results (Behera and Pattnaik, 2024).

While these advances open up great microgrid integration features, they still need to solve it on the side for cyber-security and the complication of control systems. As more and more microgrids proliferate, the research to tackle such challenges and enhance the resilience and the efficiency of microgrid systems is always being conducted.

Financial Models and Public-Private Partnerships to Support Microgrid Projects

Microgrid financial models in Which provide financing path to ensure the success and sustainability of projects for most governments: Public-private partnerships (PPPs) are joint ventures between the public and private sectors that have created a range of innovative financing mechanisms that encourage the adoption of microgrids. To solve these challenges that follow the economic development, and target the economics of enabling renewable energy generation systems in remote and underserved sites, the project financing and risk-sharing will require creative and collaborative methodologies. Through the use of public-private partnerships along with various financial frameworks, these obstacles can be overcome, resources can be pooled, risks can be shared between public and private actors, and interests can be aligned. The following pieces cover essential pieces of financial models, and PPPs that can be harnessed to help finance microgrid projects.

Microgrid Project Financial Models

Green Bonds & Investment Incentives: Similar to bonds in general, they are being widely used to finance microgrids, but green bonds are increasingly becoming a popular instrument for investment in sustainable mode energy projects as well. Unlike other types of bonds, these bonds provide a channel for the mobilization of capital specifically for environmentally beneficial projects, while attracting investors seeking sustainable development (Xu and Xu, 2024). In addition, renewable energy devices incentives and tax benefits can reduce the amount of capital cost needed at the beginning of the project starting and thus lead to give a better rate of return on project investment and in that sense make this type of microgrid projects more realistic (Husein and Chung, 2018).

Microgrids in Grid Edge Technologies: Various business models have been proposed for microgrid projects to bring profitability and scalability at microgrid levels. For example, the distribution franchisee accents allow mini-grid supervisory operators to report, amplify, and upkeep for better utility margins and domestically suitable operator return needs (Shrimali and Sen, 2020). In Colombia the business model canvas byrjj.hgh approach is used to consider distributed energy resources, contributing to sustainability and investment.

Reliability-Constrained Optimization and Hybrid-Stochastic/Robust-Based Planning Models: Financial models that incorporate risk management such as reliability-constrained optimization and hybrid-stochastic/robust based planning models are important in dealing with uncertainties of microgrid operations. These models reduce costs and increase the use of renewable energy by optimizing both the sizing and investment timing of microgrid components (Hosseinnia, Nazarpour and Talavat, 2018) (Liu *et al.*, 2021).

Public-Private Partnerships (PPPs)

Collaborative Structures: Unperturbed PPPs are a jointly composed construction that takes the NGOs and other non-inespensible interests of people. Such co-blending is an indispensable part of dispersing financial liabilities and risks relating to microgrid undertakings. Indeed, we observe the case of PPPs enhancing the improvement of electricity access in India with their direct active management on generation and including distribution(a private operator)53 optimising the power supply under policy objectives (Shrimali and Sen, 2020). Financial: Economic, Social, Environmental, Technical, and Managerial Sustainability Assessment of PPP Microgrids Specific factors are covered by the existing literature so that when considering these in a structured way, they guide ensuring the sustainability of the microgrid PPP projects in the long run.

Public support: Public buy-in is critical to success for community microgrid projects. Estimates of how much residents and businesses in the Four Corners region of the United States are willing to pay for microgrid services closely resemble PPPs that take into account social value and community preferences (Kaczmarski, 2022). There are many opportunities for microgrid projects with financial models and public-private partnerships (PPPs), but obstacles remain. These set of reasons are guaranteeing the availability of sustainable revenue streams, managing the risk associated with long-term projects, and the fair division of roles & responsibilities among the project's stakeholders. To address these challenges, new financial models and policy frameworks are needed continuously to implement conducive frameworks for engagement of the private sector and sustainable energy development.

FUTURE DIRECTIONS FOR MICROGRID DEVELOPMENT

Trends, advancements, and the necessity for sustainable energy solutions shape the future of microgrid development. Integrating advanced energy storage solutions like next generation batteries and hydrogen stores, which provide energy microgrid deployments increased flexibility and reliability, is among the key emerging trends. Also, the use of artificial intelligence (AI) and machine learning (ML) algorithms is transforming microgrid functionality by providing predictive analytics, real-time optimization, and automating decisions. This adoption is further being spurred by the growing reliance on renewable energy sources and the emergence of the hybrid microgrid system, which merges solar, wind and traditional power sources from an integration perspective, across urban, renewable energy, healthcare, agriculture, and disaster-stricken industries. Microgrid evolution is strongly dependent on digitalization and smart grid technologies. Smart sensors, IoT-enabled devices, and cloud-based platforms enable seamless monitoring, remote control, and data-driven insights, leading to operational efficiency and reduced maintenance costs. Blockchain technology also starts to play a role as a mediator, allowing peer-to-peer energy trading in microgrids, providing transparency and increasing security in the grid. Digital twin technology for microgrids builds virtual models, allowing simulation of performance, projection of potential failures, and optimization of system configurations before they are deployed. Microgrids offer a powerful avenue to achieve community energy autonomy in rural and underserved regions. Microgrids can help decrease reliance on the centralized utility, lower electricity costs, and give communities energy independence by allowing them to generate and consume energy locally. They promote economic development by generating jobs in installation, maintenance and energy management and building resilience against natural disasters or grid failures. To ensure a cleaner, sustainable future, the establishment of decentralized energy systems also benefits from government policies advocating for energy security and sustainability

CONCLUSION

Today, the energy agenda of the world is in a historic dilemma, because the double challenge of transition and resilience needs to develop new solutions that can face the increasing challenges of a climate crisis. Traditional power networks, burdened by their own heavy carbon footprint, are also increasingly vulnerable to the destructive effects of a changing climate — including more frequent blackouts, reduced efficiencies and damage to infrastructure. Thus, microgrids can be thought of as a paradigm shift in this respect, opening up pathways to cleaner, more reliable, and decentralized energy systems and in general powering communities and addressing urgent environmental issues. Microgrids have proven that the gap between energy demand and sustainability can be closed. They optimize the usage of green energy adds new renewable energy, such as solar, wind, and biomass that aid reducing fossil fuels dependency and minimize greenhouse gas emissions. They may work independently or together as evidenced by their decentralized nature, which provides a steady stream of energy during outages from the grid or disasters. This resiliency is critical in vulnerable regions where extreme weather events and failing infrastructure bring economic activity to a halt and put people's safety at risk. Besides the technical advantages, microgrids also provide socio-economic benefits in energy equity and inclusivity. They also enable energy generation and use at the source, reducing transmission losses, while offering lower-cost solutions that can effectively serve rural and underserved populations. This democratization of energy enables communities to control energy resources, promote local economic development, and improve energy security. When looking at how transportables and even traditional larger grids stand to benefit from operating experience with large scale, microscale grids are not just helping larger systems but also will turn out to be incubators and fulcrums for many breakthroughs in storage and smart grid and digital energy management systems. Although microgrids have great potential, this is not without challenges for widespread adoption. Most technical barriers, like integration with existing power systems stability, and energy flow optimization need to be solved through R& D. And there are economic challenges, including high up-front installation costs and the need for sustainable financing models, the report adds. Policy and regulatory frameworks will need to be modified to enable deployments of microgrids, encourage the integration of renewable energy, and establish clear operating standards and ownership models. In response to these challenges, a joint and multi-stakeholder industry sector approach to the common plant protection product that can be produced and used in domestic and export fields needs to be in place. An enabling environment must be created with the collaborative approach of government, industry leaders, academic institutions and local communities to incentivise the adoption of microgrids. Public-private partnerships can deliver the scale of financial and technological resources needed to tackle the latter challenge; policy measures including tax incentives, grants and subsidies can quickly deploy these technologies. It suggests that investment in education, capacity-building initiatives, is just as critical so that stakeholders have the capacity to execute and sustain microgrid systems. Insights from the Report : Microgrids: Current-state to future e versatilityKeeping current: The microgrid industry isn't standing still Emerging technologies such as artificial intelligence, blockchain, and Internet of Things (IoT) systems are also increasingly entering the fray, with the potential to enhance microgrids' efficiency, reliability, and scalability. This is because the convergence of these innovations with microgrids will be critical to building a sustainable, resilient energy future.

REFERENCES

Abbasi, M., Abbasi, E., Li, L., Aguilera, R. P., Lu, D., & Wang, F. (2023). Review on the Microgrid Concept, Structures, Components, Communication Systems, and Control Methods. *Energies*, *16*(1), 484. DOI: 10.3390/en16010484

Ahmed, I., Rehan, M., Basit, A., Ahmad, H., Ahmed, W., Ullah, N., Piecha, M., Blazek, V., & Prokop, L. (2023). Review on microgrids design and monitoring approaches for sustainable green energy networks. *Scientific Reports*, *13*(1), 21663. DOI: 10.1038/s41598-023-48985-7 PMID: 38066017

Alonso, À., de la Hoz, J., Martín, H., Coronas, S., Salas, P., & Matas, J. (2020). A Comprehensive Model for the Design of a Microgrid under Regulatory Constraints Using Synthetical Data Generation and Stochastic Optimization. *Energies*, *13*(21), 5590. DOI: 10.3390/en13215590

Awogbemi, O., Kallon, D. V. V., & Kumar, K. S. (2024). Contributions of artificial intelligence and digitization in achieving clean and affordable energy. *Intelligent Systems with Applications*, *22*, 200389. DOI: 10.1016/j.iswa.2024.200389

Bartos, M., Chester, M., Johnson, N., Gorman, B., Eisenberg, D., Linkov, I., & Bates, M. (2016). Impacts of rising air temperatures on electric transmission ampacity and peak electricity load in the United States. *Environmental Research Letters*, *11*(11), 114008. DOI: 10.1088/1748-9326/11/11/114008

Behera, P. K., & Pattnaik, M. (2024). Design and real-time implementation of wind–photovoltaic driven low voltage direct current microgrid integrated with hybrid energy storage system. *Journal of Power Sources*, *595*, 234028. DOI: 10.1016/j.jpowsour.2023.234028

Behrendt, J. (2023). Microgrids and EU law: Three Microgrid models to solve one regulatory puzzle. *Energy Policy*, *177*, 113483. DOI: 10.1016/j.enpol.2023.113483

Cabana-Jiménez, K., Candelo-Becerra, J. E., & Sousa Santos, V. (2022). Comprehensive Analysis of Microgrids Configurations and Topologies. *Sustainability (Basel)*, *14*(3), 1056. DOI: 10.3390/su14031056

Cao, W., & Zhou, L. (2024). Resilient microgrid modeling in Digital Twin considering demand response and landscape design of renewable energy. *Sustainable Energy Technologies and Assessments*, *64*, 103628. DOI: 10.1016/j.seta.2024.103628

Chen, L. (2024). Exploring Innovative Practices and Challenges in the Ecological Development of Art Industry Driven by Digital Technology. *Applied Mathematics and Nonlinear Sciences*, *9*(1), 20242975. DOI: 10.2478/amns-2024-2975

Dalir, F., Shafiepour Motlagh, M., & Ashrafi, K. (2017). Sensitivity analysis of parameters affecting carbon footprint of fossil fuel power plants based on life cycle assessment scenarios. *Global Journal of Environmental Science and Management*, *3*(1). Advance online publication. DOI: 10.22034/gjesm.2017.03.01.008

Dawson, R. J., Thompson, D., Johns, D., Wood, R., Darch, G., Chapman, L., Hughes, P. N., Watson, G. V. R., Paulson, K., Bell, S., Gosling, S. N., Powrie, W., & Hall, J. W. (2018). A systems framework for national assessment of climate risks to infrastructure. *Philosophical Transactions. Series A, Mathematical, Physical, and Engineering Sciences*, *376*(2121), 20170298. DOI: 10.1098/rsta.2017.0298 PMID: 29712793

Fan, W., Wang, Z., & Lv, W. (2023). Effects of the resilience of energy systems on economic sustainability. *Sustainable Production and Consumption*, *41*, 379–390. DOI: 10.1016/j.spc.2023.08.019

Ferrall, I., Callaway, D., & Kammen, D. M. (2022). Measuring the reliability of SDG 7: The reasons, timing, and fairness of outage distribution for household electricity access solutions. *Environmental Research Communications*, *4*(5), 055001. DOI: 10.1088/2515-7620/ac6939

Forzieri, G., Bianchi, A., Silva, F. B., Marin Herrera, M. A., Leblois, A., Lavalle, C., Aerts, J. C. J. H., & Feyen, L. (2018). Escalating impacts of climate extremes on critical infrastructures in Europe. *Global Environmental Change*, *48*, 97–107. DOI: 10.1016/j.gloenvcha.2017.11.007 PMID: 29606806

Gholami, M., Muyeen, S. M., & Abokhamis Mousavi, S. (2024). Optimal sizing of battery energy storage systems and reliability analysis under diverse regulatory frameworks in microgrids. *Energy Strategy Reviews*, *51*, 101255. DOI: 10.1016/j.esr.2023.101255

Gomez-Garcia, C. A., Askar-Rodriguez, M., & Velasco-Medina, J. (2021). Platform for Healthcare Promotion and Cardiovascular Disease Prevention. *IEEE Journal of Biomedical and Health Informatics*, *25*(7), 2758–2767. DOI: 10.1109/JBHI.2021.3051967 PMID: 33449888

Gurgel, A. C., Reilly, J., Morris, J., Schlosser, C. A., Gao, X., Yuan, M., & Tapia-Ahumada, K. (2024). Assessing compounding climate-related stresses and development pathways on the power sector in the central U.S. *Mitigation and Adaptation Strategies for Global Change*, *29*(4), 27. DOI: 10.1007/s11027-024-10119-3

Hertwich, E. G. (2021). Increased carbon footprint of materials production driven by rise in investments. *Nature Geoscience*, *14*(3), 151–155. DOI: 10.1038/s41561-021-00690-8

Hosseinnia, H., Nazarpour, D., & Talavat, V. (2018). Utilising reliability-constrained optimisation approach to model microgrid operator and private investor participation in a planning horizon. *IET Generation, Transmission & Distribution*, *12*(21), 5798–5810. DOI: 10.1049/iet-gtd.2018.5930

Husein, M., & Chung, I.-Y. (2018). Optimal design and financial feasibility of a university campus microgrid considering renewable energy incentives. *Applied Energy*, *225*, 273–289. DOI: 10.1016/j.apenergy.2018.05.036

Irmak, E., Kabalci, E., & Kabalci, Y. (2023). Digital Transformation of Microgrids: A Review of Design, Operation, Optimization, and Cybersecurity. *Energies*, *16*(12), 4590. DOI: 10.3390/en16124590

Jirdehi, M. A., Tabar, V. S., Ghassemzadeh, S., & Tohidi, S. (2020). Different aspects of microgrid management: A comprehensive review. *Journal of Energy Storage*, *30*, 101457. DOI: 10.1016/j.est.2020.101457

Kaczmarski, J. I. (2022). Public support for community microgrid services. *Energy Economics*, *115*, 106344. DOI: 10.1016/j.eneco.2022.106344

Khan Ngwashi, D., Kafatiya Arnold, A., Godwill Ndeh, S., & Tanyi, E. (2023). Optimal design and sizing of a multi-microgrids system: Case study of Goma in The Democratic Republic of the Congo. *Scientific African*, *22*, e01913. DOI: 10.1016/j.sciaf.2023.e01913

Kiehbadroudinezhad, M., Hosseinzadeh-Bandbafha, H., Rosen, M. A., Kumar Gupta, V., Peng, W., Tabatabaei, M., & Aghbashlo, M. (2023). The role of energy security and resilience in the sustainability of green microgrids: Paving the way to sustainable and clean production. *Sustainable Energy Technologies and Assessments*, *60*, 103485. DOI: 10.1016/j.seta.2023.103485

Koundouri, P., Alamanos, A., Devves, S., Landis, C., & Dellis, K. (2024). Innovations for Holistic and Sustainable Transitions. *Energies*, *17*(20), 5184. DOI: 10.3390/en17205184

Kumar, D., Bhardwaj, R., Jassal, S., Goyal, T., Khullar, A., & Gupta, N. (2021). Application of enzymes for an eco-friendly approach to textile processing. *Environmental Science and Pollution Research International*, *30*(28), 71838–71848. DOI: 10.1007/s11356-021-16764-4 PMID: 34651264

Kumar, D., & Kumar, A. (2024). 'A reliable hybrid autoregressive integrated moving average and deep reinforcement machine learning strategy for resiliency enhancement in microgrid', *Sustainable Energy. Grids and Networks*, *39*, 101424. DOI: 10.1016/j.segan.2024.101424

Lee, J.-Y., & Lin, H.-F. (2019). Multi-Footprint Constrained Energy Sector Planning. *Energies*, *12*(12), 2329. DOI: 10.3390/en12122329

Lei, Y., Ali, M., Khan, I. A., Yinling, W., & Mostafa, A. (2024). Presenting a model for decentralized operation based on the internet of things in a system multiple microgrids. *Energy*, *293*, 130637. DOI: 10.1016/j.energy.2024.130637

Li, J. (2024). Integration and development of modern art design and environmental design in the context of deep learning. *Applied Mathematics and Nonlinear Sciences*, *9*(1), 20242687. DOI: 10.2478/amns-2024-2687

Liu, J., Liang, J., Ding, J., Zhang, G., Zeng, X., Yang, Q., Zhu, B., & Gao, W. (2021). Microfiber pollution: An ongoing major environmental issue related to the sustainable development of textile and clothing industry. *Environment, Development and Sustainability*, *23*(8), 11240–11256. DOI: 10.1007/s10668-020-01173-3

Loew, A., Jaramillo, P., Zhai, H., Ali, R., Nijssen, B., Cheng, Y., & Klima, K. (2020). Fossil fuel–fired power plant operations under a changing climate. *Climatic Change*, *163*(1), 619–632. DOI: 10.1007/s10584-020-02834-y

Lopez Alzate, Y., Gómez-Luna, E., & Vasquez, J. C. (2024). Innovative Microgrid Services and Applications in Electric Grids: Enhancing Energy Management and Grid Integration. *Energies*, *17*(22), 5567. DOI: 10.3390/en17225567

Lu, Y., Xu, R., & Ma, J. (2024). Urban Energy Transition Policy Implementation: Navigating Complex Political Realities and Overcoming Barriers to Effective Strategy Deployment. *Sustainable Cities and Society*, *101*, 105127. DOI: 10.1016/j.scs.2023.105127

Ma, R., Abid, N., Yang, S., & Ahmad, F. (2023). From crisis to resilience: Strengthening climate action in OECD countries through environmental policy and energy transition. *Environmental Science and Pollution Research International*, *30*(54), 115480–115495. DOI: 10.1007/s11356-023-29970-z PMID: 37882927

Mansouri, S. A., Nematbakhsh, E., Ahmarinejad, A., Jordehi, A. R., Javadi, M. S., & Marzband, M. (2022). A hierarchical scheduling framework for resilience enhancement of decentralized renewable-based microgrids considering proactive actions and mobile units. *Renewable & Sustainable Energy Reviews*, *168*, 112854. DOI: 10.1016/j.rser.2022.112854

Morales, R., Marín, L. G., Roje, T., Caquilpan, V., Sáez, D., & Nuñez, A. (2024). Microgrid planning based on computational intelligence methods for rural communities: A case study in the José Painecura Mapuche community, Chile. *Expert Systems with Applications*, *235*, 121179. DOI: 10.1016/j.eswa.2023.121179

Mukhtiar, F. (2024). Comparing Microgrid Planning and Management Methods for Energy Resilience and Sustainability. *Power System Technology*, *48*(2), 1811–1831. DOI: 10.52783/pst.758

Neumann, J. E., Price, J., Chinowsky, P., Wright, L., Ludwig, L., Streeter, R., Jones, R., Smith, J. B., Perkins, W., Jantarasami, L., & Martinich, J. (2015). Climate change risks to US infrastructure: Impacts on roads, bridges, coastal development, and urban drainage. *Climatic Change*, *131*(1), 97–109. DOI: 10.1007/s10584-013-1037-4

Nsabimana, R., Perelman, S., Walheer, B., & Mapapa, M. (2024). Effectiveness and efficiency in access to reliable electricity: The case of East African countries. *Socio-Economic Planning Sciences*, *93*, 101898. DOI: 10.1016/j.seps.2024.101898

Onu, U. G., de Doile, G. N. D., Zambroni de Souza, A. C., & Balestrassi, P. P. (2024). Economic sustainability and social inclusion in rural electrical grid design. *Utilities Policy*, *89*, 101779. DOI: 10.1016/j.jup.2024.101779

Ottenburger, S. S., Cox, R., Chowdhury, B. H., Trybushnyi, D., Omar, E. A., Kaloti, S. A., Ufer, U., Poganietz, W.-R., Liu, W., Deines, E., Müller, T. O., Möhrle, S., & Raskob, W. (2024). Sustainable urban transformations based on integrated microgrid designs. *Nature Sustainability*, *7*(8), 1067–1079. DOI: 10.1038/s41893-024-01395-7

Parlikar, A., Tepe, B., Möller, M., Hesse, H., & Jossen, A. (2024). Quantifying the carbon footprint of energy storage applications with an energy system simulation framework — Energy System Network. *Energy Conversion and Management*, *304*, 118208. DOI: 10.1016/j.enconman.2024.118208

Ralston Fonseca, F., Craig, M., Jaramillo, P., Bergés, M., Severnini, E., Loew, A., Zhai, H., Cheng, Y., Nijssen, B., Voisin, N., & Yearsley, J. (2021). Effects of Climate Change on Capacity Expansion Decisions of an Electricity Generation Fleet in the Southeast U.S. *Environmental Science & Technology*, *55*(4), 2522–2531. DOI: 10.1021/acs.est.0c06547 PMID: 33497216

Reis, G. L., Brandao, D. I., Oliveira, J. H., Araujo, L. S., & Cardoso Filho, B. J. (2022). Case Study of Single-Controllable Microgrid: A Practical Implementation. *Energies*, *15*(17), 6400. DOI: 10.3390/en15176400

Shrimali, G., & Sen, V. (2020). Scaling reliable electricity access in India: A public-private partnership model. *Energy for Sustainable Development*, *55*, 69–81. DOI: 10.1016/j.esd.2020.01.006

Su, J., Zhang, R., Dehghanian, P., Kapourchali, M. H., Choi, S., & Ding, Z. (2024). Renewable-dominated mobility-as-a-service framework for resilience delivery in hydrogen-accommodated microgrids. *International Journal of Electrical Power & Energy Systems*, *159*, 110047. DOI: 10.1016/j.ijepes.2024.110047

Tahir, W., Farhan, M., Bhatti, A. R., Butt, A. D., & Farid, G. (2024). A modified control strategy for seamless switching of virtual synchronous generator-based inverter using frequency, phase, and voltage regulation. *International Journal of Electrical Power & Energy Systems*, *157*, 109805. DOI: 10.1016/j.ijepes.2024.109805

Tayyab, Q., Qani, N. A., Elkholy, M. H., Ahmed, S., Yona, A., & Senjyu, T. (2024). Techno-economic configuration of an optimized resident microgrid: A case study for Afghanistan. *Renewable Energy*, *224*, 120097. DOI: 10.1016/j.renene.2024.120097

Xu, J., & Xu, W. (2024). Financing sustainable smart city Projects: Public-Private partnerships and green Bonds. *Sustainable Energy Technologies and Assessments*, *64*, 103699. DOI: 10.1016/j.seta.2024.103699

Zakeri, B. (2024). Digitalization for Resilient and Sustainable Energy Transitions. *Energies*, *17*(21), 5434. DOI: 10.3390/en17215434

Zhang, Y., Cheng, C., Yang, T., Jin, X., Jia, Z., Shen, J., & Wu, X. (2022). Assessment of climate change impacts on the hydro-wind-solar energy supply system. *Renewable & Sustainable Energy Reviews*, *162*, 112480. DOI: 10.1016/j.rser.2022.112480

Zheng, L., Huang, B., Qiu, H., & Bai, H. (2024). 'The role of social media followers' agency in influencer marketing: A study based on the heuristic–systematic model of information processing'. *International Journal of Advertising*, *43*(3), 554–579. DOI: 10.1080/02650487.2023.2229148

Compilation of References

Abbasi, M., Abbasi, E., Li, L., Aguilera, R. P., Lu, D., & Wang, F. (2023). Review on the Microgrid Concept, Structures, Components, Communication Systems, and Control Methods. *Energies*, *16*(1), 484. DOI: 10.3390/en16010484

Abdulkadir, A., Ally, B., Remmen, A., Hirsbak, S., & Salukele, F. (2025). Circular Economy Framework for Plastic Waste Management: A Case Study from Coastal Hotels in Zanzibar. *Resources*, *14*(11), 175. DOI: 10.3390/resources14110175

Adhikari, S., Mahmud, M. A. P., Moon, E., & Timms, W. (2024). Comprehensive life cycle assessment of garden organic waste valorisation: A case study in regional Australia. *Journal of Cleaner Production*, *472*, 143496. Advance online publication. DOI: 10.1016/j.jclepro.2024.143496

AFAD. (2018). *Türkiye'de afet yönetimi ve doğa kaynaklı afet istatistikleri.* https://www.afad.gov.tr/kurumlar/afad.gov.tr/35429/xfiles/turkiye_de_afetler.pdf

AFAD. (2021). *Batı Karadeniz bölgesinde yaşanan sel afeti sonrasında bölgedeki çalışmalar devam ediyor.* https://www.afad.gov.tr/bati-karadeniz-bolgesinde-yasanan-sel-afeti-sonrasinda-bolgedeki-calismalar-devam-ediyor

AFAD. (2022). *AFAD hakkında.* https://www.afad.gov.tr/afad-hakkinda

Ahlborg, H., Michael, K., Unsworth, S. J., Hategekimana, S., Osunmuyiwa, O., Åberg, A., & Hultman, M. (2024). Thirty-five years of research on energy and power: A landscape analysis. *Renewable & Sustainable Energy Reviews*, *199*, 114542. Advance online publication. DOI: 10.1016/j.rser.2024.114542

Ahmad, N., & Ma, X. (2021). How does tourism development affect environmental pollution? Tourism Economics, 28(6), 1453-1479. (Original work published 2022)DOI: 10.1177/13548166211000480

Ahmed, I., Rehan, M., Basit, A., Ahmad, H., Ahmed, W., Ullah, N., Piecha, M., Blazek, V., & Prokop, L. (2023). Review on microgrids design and monitoring approaches for sustainable green energy networks. *Scientific Reports*, *13*(1), 21663. DOI: 10.1038/s41598-023-48985-7 PMID: 38066017

Akbal, A. (2023). Doğal afetler ve afet yönetiminin sosyolojik değerlendirmesi. In M. A. Kasapoğlu (Ed.), *Afet ve toplum: İnsan doğa ilişkisi* (pp. 10–23). Töz.

Alessia De Iulis. (2024, October 8). *The European Non-Governmental Sports Organisation (ENGSO).* ENGSO | the European Organisation for Grassroots Sport. https://engso.eu/sport-for-unity-resilience-and-empowerment-will-analyse-cooperation-models-focusing-on-resilience-and-inclusion/

Alonso, À., de la Hoz, J., Martín, H., Coronas, S., Salas, P., & Matas, J. (2020). A Comprehensive Model for the Design of a Microgrid under Regulatory Constraints Using Synthetical Data Generation and Stochastic Optimization. *Energies*, *13*(21), 5590. DOI: 10.3390/en13215590

Ang, P., Goh, W., Bu, J., & Cheng, S. (2024). Comparative Industrial-Scale Life Cycle Assessment of Base Case and Heat Recovery Scenarios for Carbon Capture from Natural Gas Combined Cycle Power Plants Using Aqueous Ammonia. *ACS Sustainable Chemistry & Engineering*, *12*(51), 18335–18349. DOI: 10.1021/acssuschemeng.4c06660

Anokye, K., Mohammed, A. S., Agyemang, P., Agya, B. A., Amuah, E. E. Y., & Sodoke, S. (2024). A systematic review of the impacts of open burning and open dumping of waste in Ghana: A way forward for sustainable waste management. *Cleaner Waste Systems*, *8*, 100152. Advance online publication. DOI: 10.1016/j.clwas.2024.100152

Arbues, I. C. (2021). *Understanding the feedbacks between climate change and the global economy: an ecological, post-Keynesian modelling approach for the European Green Deal*, Unpublished Master thesis, Delft University of Technology.

Aromolaran, A., & Sartaj, M. (2024). Microbial Analysis and Methane Assessment of Trinary Anaerobic Co-Digestion of Organic Fraction of Municipal Solid Waste. *BioEnergy Research*, *17*(2), 1250–1262. DOI: 10.1007/s12155-023-10623-5

ASHB. (2021a). *Bakanımız Yanık: "Başta Kastamonu olmak üzere Batı Karadeniz'i etkileyen sel felaketi 3. seviye afet ilan edilmiştir"*. https://www.aile.gov.tr/haberler/bakanimiz-yanik-basta-kastamonu-olmak-uzere-bati-karadenizi-etkileyen-sel-felaketi-ucuncu-seviye-afet-ilan-edilmistir/

ASHB. (2021b). *2021 E-bülten: Eylül–Ekim 2021*. https://www.aile.gov.tr/media/96243/bi-ri-mbultenleri-baski-1.pdf

ASHB. (2022). *2021 faaliyet raporu*. https://www.aile.gov.tr/media/100242/2021-yili-faaliyet-raporu.pdf

Awan, U., & Sroufe, R. (2022). Sustainability in the Circular Economy: Insights and Dynamics of Designing Circular Business Models. *Applied Sciences (Basel, Switzerland)*, *12*(3), 1521. DOI: 10.3390/app12031521

Awogbemi, O., Kallon, D. V. V., & Kumar, K. S. (2024). Contributions of artificial intelligence and digitization in achieving clean and affordable energy. *Intelligent Systems with Applications*, *22*, 200389. DOI: 10.1016/j.iswa.2024.200389

Axhami, M., Ndou, V., Milo, V., & Scorrano, P. (2023). Creating Value via the Circular Economy: Practices in the Tourism Sector. *Administrative Sciences*, *13*(7), 166. DOI: 10.3390/admsci13070166

Azam, M., Mahmudul Alam, M., & Haroon Hafeez, M. (2018). Md Mahmudul Alam, Muhammad Haroon Hafeez, (2018). Effect of tourism on environmental pollution: Further evidence from Malaysia, Singapore and Thailand. *Journal of Cleaner Production*, *190*, 330–338. DOI: 10.1016/j.jclepro.2018.04.168

Babalola, A. A., & Olawuyi, D. S. (2022). Overcoming Regulatory Failure in the Design and Implementation of Gas Flaring Policies: The Potential and Promise of an Energy Justice Approach. *Sustainability (Basel)*, *14*(11), 6800. Advance online publication. DOI: 10.3390/su14116800

Banaś, M., Pająk, T., Bator, J., Wróbel, W., & Ciuła, J. (2024). Energy Efficiency Analysis of Waste-to-Energy Plants in Poland. *Energies*, *17*(10), 2390. Advance online publication. DOI: 10.3390/en17102390

Bartos, M., Chester, M., Johnson, N., Gorman, B., Eisenberg, D., Linkov, I., & Bates, M. (2016). Impacts of rising air temperatures on electric transmission ampacity and peak electricity load in the United States. *Environmental Research Letters*, *11*(11), 114008. DOI: 10.1088/1748-9326/11/11/114008

Beck, U. (1992). *Risk society: Towards a new modernity* (M. Ritter, Trans.). *Sage (Atlanta, Ga.)*.

Beck, U. (2010a). Climate for change, or how to create a green modernity? *Theory, Culture & Society*, *27*(2–3), 254–266. DOI: 10.1177/0263276409358729

Beck, U. (2010b). How climate change might save the world. *Development and Society*, *39*(2), 169–183.

Behera, P. K., & Pattnaik, M. (2024). Design and real-time implementation of wind–photovoltaic driven low voltage direct current microgrid integrated with hybrid energy storage system. *Journal of Power Sources*, *595*, 234028. DOI: 10.1016/j.jpowsour.2023.234028

Behrendt, J. (2023). Microgrids and EU law: Three Microgrid models to solve one regulatory puzzle. *Energy Policy*, *177*, 113483. DOI: 10.1016/j.enpol.2023.113483

Bellezoni, R. A., Adeogun, A. P., Paes, M. X., & de Oliveira, J. A. P. (2022). Ayoola Paul Adeogun, Michel Xocaira Paes, José Antônio Puppim de Oliveira, (2022). Tackling climate change through circular economy in cities. *Journal of Cleaner Production, 381*(Part 1), 135126. DOI: 10.1016/j.jclepro.2022.135126

Bhullar, L. (2023). Meanings of energy justice in the low-carbon transition. In *Handbook of Energy Law in the Low-Carbon Transition*. De Gruyter., DOI: 10.1515/9783110752403-011

Biakhmetov, B., Li, Y., Zhao, Q., Ok, Y. S., Dostiyarov, A., Park, Y.-K., Flynn, D., & You, S. (2024). Comparing carbon-saving potential of the pyrolysis of non-recycled municipal plastic waste: Influences of system scales and end products. *Journal of Cleaner Production*, *469*, 143140. Advance online publication. DOI: 10.1016/j.jclepro.2024.143140

Bilgen, G., Balcı, E., & Kalça, M. Y. (2022). Kastamonu Bozkurt ilçesinde 11.08.2021 tarihinde meydana gelen sel felaketinin yerinde incelenmesi. *Tasarım Mimarlık ve Mühendislik Dergisi*, *2*(1), 20–35. https://izlik.org/JA36EX57DJ

Bozkurt Municipality. (2010). *Bozkurt: Güzellik onun doğasında*. Asır.

Brandt, M. (2022). Nature Lovers Don't Equal Sustainable Travelers, https://www.statista.com/chart/27611/survey-on-the-importance-of-nature-and-sustainability-when-travelling/?srsltid=AfmBOopxJjqKJcO4A_Fu05VrlnraT9J2L7pxM66wn0pL598SAd-8bSKk

Brennen, J. S., & Kreiss, D. (2014). Digitalization and digitization. *Culture Digitally*. 8(2), 10-19. https://culturedigitally.org/2014/09/digitalization-and-digitization/

Brennen, J. S., & Kreiss, D. (2016). Digitalization. In *The International Encyclopedia of Communication Theory and Philosophy*. 1-11. DOI: 10.1002/9781118766804.wbiect111

Bullard, R. D. (2000). *Dumping in Dixie: Race, class, and environmental quality* (3rd ed.). Westview Press.

Bullard, R. D. (2001). Environmental justice in the 21st century: Race still matters. *Phylon, 49*(3/4), 151–171. DOI: 10.2307/3132626

Bullard, R., & Wright, B. H. (1987). Environmentalism and the politics of equity: Emergent trends in the Black community. *Social Thought & Research, 12*(2), 21–38. DOI: 10.17161/STR.1808.5017

Cabana-Jiménez, K., Candelo-Becerra, J. E., & Sousa Santos, V. (2022). Comprehensive Analysis of Microgrids Configurations and Topologies. *Sustainability (Basel), 14*(3), 1056. DOI: 10.3390/su14031056

Camilleri, M. A. (2021). Sustainable Production and Consumption of Food. Mise-en-Place Circular Economy Policies and Waste Management Practices in Tourism Cities. *Sustainability (Basel), 13*(17), 9986. DOI: 10.3390/su13179986

Cao, W., & Zhou, L. (2024). Resilient microgrid modeling in Digital Twin considering demand response and landscape design of renewable energy. *Sustainable Energy Technologies and Assessments, 64*, 103628. DOI: 10.1016/j.seta.2024.103628

Çevre, T. C. Şehircilik ve İklim Değişikliği Bakanlığı, (2024). Sıfır Atık ile geri kazanım oranı % 35'e ulaştı., https://cygm.csb.gov.tr/sifir-atik-ile-geri-kazanim-orani-35e-ulasti.-haber-286897

Chen, L. (2024). Exploring Innovative Practices and Challenges in the Ecological Development of Art Industry Driven by Digital Technology. *Applied Mathematics and Nonlinear Sciences, 9*(1), 20242975. DOI: 10.2478/amns-2024-2975

Chen, X. (2022). Machine learning approach for a circular economy with waste recycling in smart cities. *Energy Reports, 8*, 3127–3140. DOI: 10.1016/j.egyr.2022.01.193

Chiwaridzo, O. T., & Han, J. (2025). Circular economy strategies for waste management: The mediating role of green tourism supply chain management in environmental management and energy sustainability within the Zimbabwean tourism sector. Waste Management & Research. *Waste Management & Research, 0*(0), 0734242X251385933. Advance online publication. DOI: 10.1177/0734242X251385933 PMID: 41217274

Cho, Y., Withana, P. A., Rhee, J. H., Lim, S. T., Lim, J. Y., Park, S. W., & Ok, Y. S. (2024). Achieving the sustainable waste management of medical plastic packaging using a life cycle assessment approach. *Heliyon, 10*(19), e38185. Advance online publication. DOI: 10.1016/j.heliyon.2024.e38185 PMID: 39398058

Christis, M., Athanassiadis, A., & Vercalsteren, A. (2019). Aristide Athanassiadis, An Vercalsteren, (2019). Implementation at a city level of circular economy strategies and climate change mitigation – the case of Brussels. *Journal of Cleaner Production, 218*, 511–520. DOI: 10.1016/j.jclepro.2019.01.180

Cordner, A. (2021). Risk. In Caniglia, B. S. (Eds.), *Handbook of environmental sociology* (pp. 167–187). Springer., DOI: 10.1007/978-3-030-77712-8_9

Coskuner, G., Jassim, M. S., Jaffer, S., & Ali, S. (2021). Greenhouse Gas Mitigation and Energy Recovery Potential for Optimum Waste Management Scenarios in Bahrain. 2021 3rd International Sustainability and Resilience Conference: Climate Change, 181 – 185. DOI: 10.1109/IEEECONF53624.2021.9668111

Dafermos, Y., Nikolaidi, M., & Galanis, G. (2018). Climate change, financial stability and monetary policy. *Ecological Economics, 152*, 219–234. DOI: 10.1016/j.ecolecon.2018.05.011

Dalir, F., Shafiepour Motlagh, M., & Ashrafi, K. (2017). Sensitivity analysis of parameters affecting carbon footprint of fossil fuel power plants based on life cycle assessment scenarios. *Global Journal of Environmental Science and Management, 3*(1). Advance online publication. DOI: 10.22034/gjesm.2017.03.01.008

Dauvergne, P. (2009). *Historical dictionary of environmentalism*. Scarecrow Press.

Dawson, R. J., Thompson, D., Johns, D., Wood, R., Darch, G., Chapman, L., Hughes, P. N., Watson, G. V. R., Paulson, K., Bell, S., Gosling, S. N., Powrie, W., & Hall, J. W. (2018). A systems framework for national assessment of climate risks to infrastructure. *Philosophical Transactions. Series A, Mathematical, Physical, and Engineering Sciences*, *376*(2121), 20170298. DOI: 10.1098/rsta.2017.0298 PMID: 29712793

Demirbilek, M. (2016). Environment, environmental refugees and green social work. *International Journal of Social Sciences and Education Research*, *2*(3), 905–914. DOI: 10.24289/ijsser.279033

Des Marais, E. A., Bhadra, S., & Dyer, A. R. (2012). In the wake of Japan's triple disaster: Rebuilding capacity through international collaboration. *Advances in Social Work*, *13*(2), 340–357. DOI: 10.18060/1964

Destrée, P. (2024). Gaseous politics: Contradictions and moral frontiers of the energy transition in Ghana. *Critique of Anthropology*, *44*(3), 235–255. DOI: 10.1177/0308275X241269579

Ding, Z., Wang, X., & Patrick, X. W. (2023). Zou, (2023). Barriers and countermeasures of construction and demolition waste recycling enterprises under circular economy. *Journal of Cleaner Production*, *420*, 138235. DOI: 10.1016/j.jclepro.2023.138235

Directorate of Climate Change (2024). *Paris Anlaşması*. https://iklim.gov.tr/paris-anlasmasi-i-34

Disaster Management in Japan. (2022). *Consecutive disasters: Toward the establishment of a disaster-conscious society*. https://www.bousai.go.jp/en/documentation/white_paper/pdf/SF1-1.pdf

Dominelli, L. (2015). Greening social work. In *International encyclopedia of social and behavioral sciences* (2nd ed.). DOI: 10.1016/B978-0-08-097086-8.28037-9

Dominelli, L. (2012). *Green social work: From environmental crises to environmental justice*. Polity Press.

Dominelli, L. (2014). Promoting environmental justice through green social work practice: A key challenge for practitioners and educators. *International Social Work, 57*(4), 338–345. DOI: 10.1177/0020872814524968

Dominelli, L. (2024). *Social work practice during times of disaster: A transformative green social work model*. Routledge.

Dong, Y., Zhang, Y., & Liu, S. (2024). The impacts and instruments of energy transition regulations on environmental pollution. *Environmental Impact Assessment Review, 105*, 107448. Advance online publication. DOI: 10.1016/j.eiar.2024.107448

EFL. (n.d.). *EFL Green Clubs*. EFL. https://www.efl.com/about-the-efl/green-clubs/

EM-DAT. (2023). *2022 disasters in numbers*. https://www.cred.be/sites/default/files/2022_EMDAT_report.pdf

EM-DAT. (2024a). *Disaster year in 2023* (Issue No. 74). https://files.emdat.be/2024/04/CredCrunch74.pdf

EM-DAT. (2024b). *EM-DAT overview*. https://www.emdat.be

Encyclopedia of Philosophy. (2024). *The American environmental justice movement*. https://iep.utm.edu/enviro-j/

EPA (2024). *Learn about environmental justice*. https://www.epa.gov/environmentaljustice/learn-about-environmental-justice

Erickson, C. L. (2018). *Environmental justice as social work practice*. Oxford University Press.

Euronews (2021). *Sosyal adalet endeksi*. https://tr.euronews.com/2021/11/17/sosyal-adalet-endeksi-turkiye-ab-ve-oecd-ulkeleri-aras-nda-sondan-2-s-rada

European Commission. (2015). *Sustainable Development Goals.* https://international-partnerships.ec.europa.eu/policies/sustainable-development-goals_en

European Commission. (2019). *European Green Deal. Striving to be the first climate-neutral continent.* https://commission.europa.eu/strategy-and-policy/priorities-2019-2024/european-green-deal_en

European Commission. (2020). *Recovery plan for Europe.* https://commission.europa.eu/strategy-and-policy/recovery-plan-europe_en

European Commission. (2021a). *Delivering the European Green Deal.* https://commission.europa.eu/strategy-and-policy/priorities-2019-2024/european-green-deal/delivering-european-green-deal_en

European Commission. (2021b). *European Climate Law.* https://climate.ec.europa.eu/eu-action/european-climate-law_en

European Commission. (2023). *Communication from the Commission to the European Parliament, the European Council, the Council, the European Economic and Social Committee and the Committee of the Regions. A Green Deal Industrial Plan for the Net-Zero Age.* https://eur-lex.europa.eu/legal-content/EN/TXT/PDF/?uri=CELEX:52023DC0062

European Commission. (2024). *Shaping Europe's digital future. DESI indicators.* https://digital-decade-desi.digital-strategy.ec.europa.eu/datasets/desi/charts/desi-indicators?period=desi_2024&indicator=desi_ai&breakdown=ent_all_xfin&unit=pc_ent&country=AT,BE,BG,HR,CY,CZ,DK,EE,EU,FI,FR,DE,EL,HU,IE,IT,LV,LT,LU,MT,NL,PL,PT,RO,SK,SI,ES,S

European Court of Auditors. (2023). *Special report: Circular Economy - Slow Transition by Member States despite EU action.* https://www.eca.europa.eu/en/publications/sr-2023-17

European Investment Bank, (2023). The EIB Circular Economy Guide: Supporting the circular transition – May 2023.

European parliament. (2021). *Report on artificial intelligence in criminal law and its use by the police and judicial authorities in criminal matters.* https://www.europarl.europa.eu/doceo/document/A-9-2021-0232_EN.html

European Parliament. (2022, June 22). *Green Deal: key to a climate-neutral and sustainable EU.* European Parliament. https://www.europarl.europa.eu/topics/en/article/20200618STO81513/green-deal-key-to-a-climate-neutral-and-sustainable-eu

European Parliament. (2023). Circular economy: definition, importance and benefits, https://www.europarl.europa.eu/topics/en/article/20151201STO05603/circular-economy-definition-importance-and-benefits

Eurostat. (2024a). *Circular material use rate.* DOI: 10.2908/CEI_SRM030

Eurostat. (2024b). *Consumption footprint.* DOI: 10.2908/CEI_GSR010

Eurostat. (2024c). Greenhouse gases emissions from production activities.

Eurostat. (2024d). *Persons employed in circular economy sectors.* DOI: 10.2908/CEI_CIE011

Eurostat. (2024e). *Population on 1 January by age and sex.* DOI: 10.2908/DEMO_PJAN

Eurostat. (2024f). *Waste generation per capita.* DOI: 10.2908/CEI_PC034

Ezz, H., Ibrahim, M. G., Fujii, M., & Nasr, M. (2024). Sustainable management of petrochemical wastewater using algal-bacterial granules followed by biogas and biochar production: A techno-economic perspective. *Journal of Water Process Engineering*, *68*, 106391. Advance online publication. DOI: 10.1016/j.jwpe.2024.106391

Falcone, P. M. (2019). Tourism-Based Circular Economy in Salento (South Italy): A SWOT-ANP Analysis. *Social Sciences (Basel, Switzerland)*, *8*(7), 216. DOI: 10.3390/socsci8070216

Fang, M., Njangang, H., Padhan, H., Simo, C., & Yan, C. (2023). Social media and energy justice: A global evidence. *Energy Economics, 125*, 106886. Advance online publication. DOI: 10.1016/j.eneco.2023.106886

Fan, W., Wang, Z., & Lv, W. (2023). Effects of the resilience of energy systems on economic sustainability. *Sustainable Production and Consumption, 41*, 379–390. DOI: 10.1016/j.spc.2023.08.019

Ferrall, I., Callaway, D., & Kammen, D. M. (2022). Measuring the reliability of SDG 7: The reasons, timing, and fairness of outage distribution for household electricity access solutions. *Environmental Research Communications*, *4*(5), 055001. DOI: 10.1088/2515-7620/ac6939

FIFA. (2022). *Sustainability: Waste management at tournament sites.* FIFA Publications. https://publications.fifa.com/en/final-sustainability-report/environmental-pillar/waste-minimisation/waste-management-at-tournament-sites/

Filipova, M., Gardan, I., Bondrea, A., & Gardan, D. (2024). Relationship between organizational culture and corporate sustainability. *Annals of Spiru Haret University. Economic Series*, *24*(1), 19–40. DOI: 10.26458/2411

Fontana, G., & Sawyer, M. (2016). Towards post-Keynesian ecological macroeconomics. *Ecological Economics*, *121*, 186–195. DOI: 10.1016/j.ecolecon.2015.03.017

Forzieri, G., Bianchi, A., Silva, F. B., Marin Herrera, M. A., Leblois, A., Lavalle, C., Aerts, J. C. J. H., & Feyen, L. (2018). Escalating impacts of climate extremes on critical infrastructures in Europe. *Global Environmental Change*, *48*, 97–107. DOI: 10.1016/j.gloenvcha.2017.11.007 PMID: 29606806

Francis, A. E., Webb, M., Desha, C., Rundle-Thiele, S., & Caldera, S. (2023). Environmental Sustainability in Stadium Design and Construction: A Systematic Literature Review. *Sustainability (Basel)*, *15*(8), 6896. DOI: 10.3390/su15086896

Frenzel, A., Muench, J. C., Bruckner, M., & Veit, D. (2021). Digitization or digitalization?-Toward an understanding of definitions, use and application in IS research. *Amcis 2021 Proceedings, 18.* Montreal. https://aisel.aisnet.org/amcis2021/adv_info_systems_general_track/adv_info_systems_general_track/18

Galbraith, J. K. (2021). Economics and the climate catastrophe. *Globalizations*, *18*(7), 1117–1122. DOI: 10.1080/14747731.2020.1807858

Gamage, A., Gangahagedara, R., Subasinghe, S., Gamage, J., Guruge, C., Senaratne, S., Randika, T., Rathnayake, C., Hameed, Z., Madhujith, T., & Merah, O. (2024). Advancing sustainability: The impact of emerging technologies in agriculture. *Current Plant Biology*, *40*, 100420. DOI: 10.1016/j.cpb.2024.100420

Gao, J., & Zhang, L. (2021). Exploring the dynamic linkages between tourism growth and environmental pollution: New evidence from the Mediterranean countries. *Current Issues in Tourism*, *24*(1), 49–65. DOI: 10.1080/13683500.2019.1688767

Geneva Environment Network. (2023, June 1). *Sports, the Environment, and the Role of International Geneva*. Www.genevaenvironmentnetwork.org. https://www.genevaenvironmentnetwork.org/resources/updates/sports-the-environment-and-the-role-of-international-geneva/

Ghamari, M., Sun, D., Dai, Y., See, C. H., Yu, H., Edirisinghe, M., & Sundaram, S. (2024). Valorization of diverse waste-derived nanocellulose for multifaceted applications: A review. *International Journal of Biological Macromolecules*, *280*, 136130. Advance online publication. DOI: 10.1016/j.ijbiomac.2024.136130 PMID: 39443179

Gholami, M., Muyeen, S. M., & Abokhamis Mousavi, S. (2024). Optimal sizing of battery energy storage systems and reliability analysis under diverse regulatory frameworks in microgrids. *Energy Strategy Reviews*, *51*, 101255. DOI: 10.1016/j.esr.2023.101255

Giurea, R., Precazzini, I., Ionescu, G., Ragazzi, M., & Schiavon, M. (2022, August 17). Ilaria Precazzini, Gabriela Ionescu, Marco Ragazzi, Marco Schiavon, (2022). Circular economy, waste and energy management for a sustainable agro-tourism. *AIP Conference Proceedings*, *2437*(1), 020097. DOI: 10.1063/5.0093290

Gomez-Garcia, C. A., Askar-Rodriguez, M., & Velasco-Medina, J. (2021). Platform for Healthcare Promotion and Cardiovascular Disease Prevention. *IEEE Journal of Biomedical and Health Informatics*, *25*(7), 2758–2767. DOI: 10.1109/JBHI.2021.3051967 PMID: 33449888

Görgün Baran, A. (2023). Depremin medyadaki yansımaları. In M. A. Kasapoğlu (Ed.), *Afet ve toplum: İnsan doğa ilişkisi* (pp. 83–92). Töz.

Gurgel, A. C., Reilly, J., Morris, J., Schlosser, C. A., Gao, X., Yuan, M., & Tapia-Ahumada, K. (2024). Assessing compounding climate-related stresses and development pathways on the power sector in the central U.S. *Mitigation and Adaptation Strategies for Global Change*, *29*(4), 27. DOI: 10.1007/s11027-024-10119-3

Hai, T., Alenizi, F. A., Alshahri, A. H., Chauhan, B. S., Metwally, A. S. M., & Almujibah, H. R. (2023). Energy and environmental analyses of a sustainable multi-generation municipal solid waste-to-energy integrated system for hydrogen production. *Process Safety and Environmental Protection*, *177*, 307–321. DOI: 10.1016/j.psep.2023.06.030

Haldar, S., Peddibhotla, A., & Bazaz, A. (2023). Analysing intersections of justice with energy transitions in India - A systematic literature review. *Energy Research & Social Science*, *98*, 103010. Advance online publication. DOI: 10.1016/j.erss.2023.103010

Heffron, R. J. (2024). Energy justice–the triumvirate of tenets revisited and revised. *Journal of Energy & Natural Resources Law*, *42*(2), 227–233. DOI: 10.1080/02646811.2023.2256593

Heffron, R. J., & Sokołowski, M. M. (2024). Resolving energy policy failure: Introducing energy justice as the solution to achieve a just transition. *Energy Policy*, *187*, 114042. Advance online publication. DOI: 10.1016/j.enpol.2024.114042

Heise, A. (2022). *Growth and sustainability in post- Keynesian perspective: Some notes*. ZÖSS-Discussion Papers.

Hertwich, E. G. (2021). Increased carbon footprint of materials production driven by rise in investments. *Nature Geoscience*, *14*(3), 151–155. DOI: 10.1038/s41561-021-00690-8

Hosseinnia, H., Nazarpour, D., & Talavat, V. (2018). Utilising reliability-constrained optimisation approach to model microgrid operator and private investor participation in a planning horizon. *IET Generation, Transmission & Distribution*, *12*(21), 5798–5810. DOI: 10.1049/iet-gtd.2018.5930

HOW THE 2020 TOKYO OLYMPICS SUSTAINABILITY AND ENVIRONMENTAL AWARENESS. (2020). Gamers. https://vocal.media/gamers/how-the-2020-tokyo-olympics-sustainability-and-environmental-awareness

Huang, X. (2024). The Promotion of Anaerobic Digestion Technology Upgrades in Waste Stream Treatment Plants for Circular Economy in the Context of "Dual Carbon": Global Status, Development Trend, and Future Challenges. *Water (Basel)*, *16*(24), 3718. Advance online publication. DOI: 10.3390/w16243718

Husein, M., & Chung, I.-Y. (2018). Optimal design and financial feasibility of a university campus microgrid considering renewable energy incentives. *Applied Energy*, *225*, 273–289. DOI: 10.1016/j.apenergy.2018.05.036

Huwe, D., & Rehm, M. (2022). The ecological crisis and post-Keynesian economics – bridging the gap? *European Journal of Economics and Economic Policies: Intervention*, *19*(3), 397–414.

IASC. (2007). *Guidelines on mental health and psychosocial support in emergency settings*. https://interagencystandingcommittee.org/sites/default/files/migrated/2020-11/IASC Guidelines on Mental Health and Psychosocial Support in Emergency Settings %28English%29.pdf

Ibarra-Esparza, F. E., González-López, M. E., Ibarra-Esparza, J., Lara-Topete, G. O., Senés-Guerrero, C., Cansdale, A., Forrester, S., Chong, J. P. J., & Gradilla-Hernández, M. S. (2023). Implementation of anaerobic digestion for valorizing the organic fraction of municipal solid waste in developing countries: Technical insights from a systematic review. *Journal of Environmental Management*, *347*, 118993. Advance online publication. DOI: 10.1016/j.jenvman.2023.118993 PMID: 37751665

Ibrahim, R. A., Inan, H., & Fahim, I. S. (2023). A comparative cradle-to-gate life cycle assessment of three cotton stalk waste sustainable applications. *Scientific Reports*, *13*(1), 20781. Advance online publication. DOI: 10.1038/s41598-023-47817-y PMID: 38012270

IFSW (2020). *Global standards for social work education and training*. https://www.ifsw.org/global-standards-for-social-work-education-and-training/

INFORM (2025). *Results and data*. https://drmkc-jrc-ec-europa-eu.translate.goog/inform-index/INFORM-Risk/Results-and-data

Inside FIFA. (2023). Fifa.com. https://inside.fifa.com/social-impact/sustainability/fwc-2026-strategy/strategy

International Energy Agency (2023). "Energy Access Report."

International Energy Agency. (2024, August 2). *Greenhouse Gas Emissions from Energy Data Explorer – Data Tools*. IEA. https://www.iea.org/data-and-statistics/data-tools/greenhouse-gas-emissions-from-energy-data-explorer

International Environmental Technology. (2018). *What Recycling Measures Are in Place at the World Cup?* Envirotech Online; Environmental. https://www.envirotech-online.com/news/business-news/44/breaking-news/what-recycling-measures-are-in-place-at-the-world-cup/46570

International Olympic Committee. (2020). *Requirements to host the Olympic Games*. Olympics.com. https://olympics.com/ioc/becoming-an-olympic-games-host/requirements-to-host-the-olympic-games

IOC. (2021, May 17). *Olympic Agenda 2020 - Strategic Roadmap for the Olympic Movement*. International Olympic Committee. https://olympics.com/ioc/olympic-agenda-2020

IPCC (2007). Climate Change 2007: *Synthesis report*. https://www.ipcc.ch/site/assets/uploads/2018/02/ar4_syr_full_report.pdf

Irmak, E., Kabalci, E., & Kabalci, Y. (2023). Digital Transformation of Microgrids: A Review of Design, Operation, Optimization, and Cybersecurity. *Energies*, *16*(12), 4590. DOI: 10.3390/en16124590

Islam, N. F., Gogoi, B., Saikia, R., Yousaf, B., Narayan, M., & Sarma, H. (2024). Bhoirob Gogoi, Rimon Saikia, Balal Yousaf, Mahesh Narayan, Hemen Sarma, (2024). Encouraging circular economy and sustainable environmental practices by addressing waste management and biomass energy production. *Regional Sustainability*, *5*(4), 100174. DOI: 10.1016/j.regsus.2024.100174

Japan, C. W. S. (2019). Towards mabi's recovery: lessons on year on. Retrieved from: https://www.cwsjapan.org/wp-content/uploads/2019/09/TowardsMabisRecovery_S.pdf

Japan, C. W. S. (2023). Lessons from Mabi: Five years of recovery form the 2018 Western Japan Flood. https://www.preventionweb.net/publication/lessons-mabi-five-years-recovery-2018-western-japan-flood

Japanese Red Cross Society. (2018). *Response to Japan floods and landslides 2018*. https://www.jrc.or.jp/english/relief/180724_005361.html

Jeetah, P., Somaroo, G., Surroop, D., Ragen, A. K., & Amode, N. S. (2024). Adoption of Machine Learning for Sustainable Solid Waste Management. In *Artificial Intelligence, Engineering Systems and Sustainable Development: Driving the UN SDGs*. Emerald Group Publishing Ltd., DOI: 10.1108/978-1-83753-540-820241002

Jessen, J. (2024, August 29). *F1 Accelerates Sustainability in Partnership with Aggreko*. Sustainabilitymag.com; Bizclik Media Ltd. https://sustainabilitymag.com/articles/f1-accelerates-sustainability-in-partnership-with-aggreko

Jirdehi, M. A., Tabar, V. S., Ghassemzadeh, S., & Tohidi, S. (2020). Different aspects of microgrid management: A comprehensive review. *Journal of Energy Storage, 30*, 101457. DOI: 10.1016/j.est.2020.101457

Journal of Population and Social Studies. (2017). *Sustainable Solid Waste Management in Sports Events: A Case Study of Football Matches in Thailand.* Retrieved from https://so03.tci-thaijo.org/index.php/jpss/article/view/102290

Kabil, M., Rahmat, A. F., Hegedüs, M., & Dávid, L. D. (2024). Circular Economy and Tourism: A Bibliometric Journey Through Scholarly Discourse. *Journal of Circular Economy, 2*(1). Advance online publication. DOI: 10.55845/HGWO7144

Kaczmarski, J. I. (2022). Public support for community microgrid services. *Energy Economics, 115*, 106344. DOI: 10.1016/j.eneco.2022.106344

Kalair, A. R., Seyedmahmoudian, M., Stojcevski, A., Abas, N., & Khan, N. (2021). Waste to energy conversion for a sustainable future. *Heliyon, 7*(10), e08155. DOI: 10.1016/j.heliyon.2021.e08155 PMID: 34729426

Karagüzel, , R., Türkkan Karaoğlu, G., Yılmaz, K., K. Çatlıoğlu, B., Çan, T & Akay, S., S. (2021). Kastamonu-Bozkurt 11 Ağustos 2021 taşkını ve düşündürdükleri. *Yer Mühendisliği, 16*(October), 14–25.

Karak, S. Garima, Berry, E., & Choudhary, A. K. (2024). Heavy Metal Waste Management to Combat Climate Crisis: An Overview of Plant-Based Strategies and Its Current Developments. In Integrated Waste Management: A Sustainable Approach from Waste to Wealth. Springer Singapore. DOI: 10.1007/978-981-97-0823-9_9

Karaosmanoğlu, A. (2025). Türkiye'nin 10 yıllık "Ulusal Atık Yönetimi Stratejisi ve Planı" hazırlandı, https://www.aa.com.tr/tr/ekonomi/turkiyenin-10-yillik-ulusal-atik-yonetimi-stratejisi-ve-plani-hazirlandi/3719383
Khan, S., & Haleem, A. (2021). Investigation of circular economy practices in the context of emerging economies: a CoCoSo approach. International Journal of Sustainable Engineering, 14(3), 357–367. DOI: 10.1080/19397038.2020.1871442

Karataş, K. (2023). *Afet yaşamış toplumlarla çalışma* [Video]. YouTube. https://www.youtube.com/watch?v=cNQSHhPnldc

Kaswan, A. (2023). Background on environmental justice and racism. In *The Palgrave Handbook of Global Sustainability* (Vol. 2–3). Springer International Publishing., DOI: 10.1007/978-3-031-01949-4_70

Katircioglu, S. T. (2014). International tourism, energy consumption, and environmental pollution: The case of Turkey. *Renewable & Sustainable Energy Reviews*, *36*, 180–187. DOI: 10.1016/j.rser.2014.04.058

Kaur, A., & Sharma, V. (2024). Nexus between climate change and mitigation approaches for sustainable development: A bibliometric review. *Global Knowledge, Memory and Communication*. Advance online publication. DOI: 10.1108/GKMC-11-2023-0418

Kaur, H., Singh, K., Kumar, P., & Kaur, A. (2023). Assessing the environmental sustainability corridor: An empirical study of Renewable energy consumption in BRICS nation. *IOP Conference Series. Earth and Environmental Science*, *1110*(1), 012053. Advance online publication. DOI: 10.1088/1755-1315/1110/1/012053

Kazazi, M. (2024, December 18). *Sustainable Training Methods for Athletes*. The Sports Financial Literacy Academy. https://moneysmartathlete.com/athletes-sustainable-practices-in-sports/sustainable-training-methods-for-athletes/

Keen, S. (2021). The appallingly bad neoclassical economics of climate change. *Globalizations*, *18*(7), 1149–1177. DOI: 10.1080/14747731.2020.1807856

Kemp-Benedict, E. (2018). Energy, growth, and sustainability: Five propositions. *Ecological Economics*, *154*, 78–83.

Khalid, W., & Rogstam, J. (2013). Energy usage prediction model comparing outdoor vs. indoor ice rinks. *Energy and Building*, *67*, 195–200. DOI: 10.1016/j.enbuild.2013.08.004

Khan Ngwashi, D., Kafatiya Arnold, A., Godwill Ndeh, S., & Tanyi, E. (2023). Optimal design and sizing of a multi-microgrids system: Case study of Goma in The Democratic Republic of the Congo. *Scientific African*, *22*, e01913. DOI: 10.1016/j.sciaf.2023.e01913

Kiehbadroudinezhad, M., Hosseinzadeh-Bandbafha, H., Rosen, M. A., Kumar Gupta, V., Peng, W., Tabatabaei, M., & Aghbashlo, M. (2023). The role of energy security and resilience in the sustainability of green microgrids: Paving the way to sustainable and clean production. *Sustainable Energy Technologies and Assessments*, *60*, 103485. DOI: 10.1016/j.seta.2023.103485

Klein, N., Ramos, T. B., & Deutz, P. (2020). Circular Economy Practices and Strategies in Public Sector Organizations: An Integrative Review. *Sustainability (Basel)*, *12*(10), 4181. DOI: 10.3390/su12104181

Klitgaard, K., & Krall, L. (2012). Ecological economics, degrowth, and institutional change. *Ecological Economics*, *84*, 247–253. DOI: 10.1016/j.ecolecon.2011.11.008

Koundouri, P., Alamanos, A., Devves, S., Landis, C., & Dellis, K. (2024). Innovations for Holistic and Sustainable Transitions. *Energies*, *17*(20), 5184. DOI: 10.3390/en17205184

Kumar, D., Bhardwaj, R., Jassal, S., Goyal, T., Khullar, A., & Gupta, N. (2021). Application of enzymes for an eco-friendly approach to textile processing. *Environmental Science and Pollution Research International*, *30*(28), 71838–71848. DOI: 10.1007/s11356-021-16764-4 PMID: 34651264

Kumar, D., & Kumar, A. (2024). 'A reliable hybrid autoregressive integrated moving average and deep reinforcement machine learning strategy for resiliency enhancement in microgrid', *Sustainable Energy. Grids and Networks*, *39*, 101424. DOI: 10.1016/j.segan.2024.101424

Laureti, L., Costantiello, A., Anobile, F., Leogrande, A., & Magazzino, C. (2024). Waste Management and Innovation: Insights from Europe. *Recycling*, *9*(5), 82. DOI: 10.3390/recycling9050082

Lavier, T. (2024, July 25). *ITTF Welcomes Paris Agreement for Sport and Sustainable Development - International Table Tennis Federation.* International Table Tennis Federation. https://www.ittf.com/2024/07/25/ittf-welcomes-paris-agreement-for-sport-and-sustainable-development/

Lee, J.-Y., & Lin, H.-F. (2019). Multi-Footprint Constrained Energy Sector Planning. *Energies*, *12*(12), 2329. DOI: 10.3390/en12122329

Legner, C., Eymann, T., Hess, T., Matt, C., Böhmann, T., Drews, P., & Ahlemann, F. (2017). Digitalization: Opportunity and challenge for the business and information systems engineering community. *Business & Information Systems Engineering*, *59*(4), 301–308. DOI: 10.1007/s12599-017-0484-2

Lei, Y., Ali, M., Khan, I. A., Yinling, W., & Mostafa, A. (2024). Presenting a model for decentralized operation based on the internet of things in a system multiple microgrids. *Energy*, *293*, 130637. DOI: 10.1016/j.energy.2024.130637

Li, J. (2024). Integration and development of modern art design and environmental design in the context of deep learning. *Applied Mathematics and Nonlinear Sciences*, *9*(1), 20242687. DOI: 10.2478/amns-2024-2687

Lin, S.-S., Zhang, N., Xu, Y.-S., & Hino, T. (2020). Lessons learned from catastrophic floods in western Japan in 2018: Sustainable perspective analysis. *Water (Basel)*, *12*(9), 1-15. DOI: 10.3390/w12092489

Liu, S. (2022a). Excess waste a key sustainability concern for travelers, https://www.statista.com/chart/28149/excess-waste-a-key-sustainability-concern/?srsltid=AfmBOopWKGMdBlm-QyBp1jdFIxj165U2rb2TDaCULJ5adtTpggFuNPTl

Liu, S. (2022b). Consumers want to travel more sustainably, https://www.statista.com/chart/27980/travel-more-sustainably/?srsltid=AfmBOoqU2vAfdpFurB6nUF-_f4NHdn0VznW6ep5edaSX_tuOuG_cKpat

Liu, J., Liang, J., Ding, J., Zhang, G., Zeng, X., Yang, Q., Zhu, B., & Gao, W. (2021). Microfiber pollution: An ongoing major environmental issue related to the sustainable development of textile and clothing industry. *Environment, Development and Sustainability*, *23*(8), 11240–11256. DOI: 10.1007/s10668-020-01173-3

Liu, J., Wu, P., Jiang, Y., & Wang, X. (2021). Explore potential barriers of applying circular economy in construction and demolition waste recycling. *Journal of Cleaner Production*, *326*, 129400. DOI: 10.1016/j.jclepro.2021.129400

LiveScience. (2011). *Japan's explosive geology explained.* https://www.livescience.com/30226-japan-tectonics-explosive-geology-ring-of-fire-110314.html

Loew, A., Jaramillo, P., Zhai, H., Ali, R., Nijssen, B., Cheng, Y., & Klima, K. (2020). Fossil fuel–fired power plant operations under a changing climate. *Climatic Change*, *163*(1), 619–632. DOI: 10.1007/s10584-020-02834-y

Lokesh Bohra. (2024, December 27). *Sustainable Play: Navigating ESG Trends in the Sports Industry*. Medium. https://medium.com/@lokeshbohra/sustainable-play-navigating-esg-trends-in-the-sports-industry-december-2024-9cbb7490fdf6

Longfor, N. R., Aduba, J., Istrate, I.-R., & Qian, X. (2023). Mitigating greenhouse gas emissions from municipal solid waste in Sub-Saharan Africa via sustainable waste management: An economic benefit assessment. *Resources, Conservation & Recycling Advances*, *20*, 200192. Advance online publication. DOI: 10.1016/j.rcradv.2023.200192

Lopez Alzate, Y., Gómez-Luna, E., & Vasquez, J. C. (2024). Innovative Microgrid Services and Applications in Electric Grids: Enhancing Energy Management and Grid Integration. *Energies*, *17*(22), 5567. DOI: 10.3390/en17225567

Löscher, A., & Kaltenbrunner, A. (2023). Climate change and macroeconomic policy space in developing and emerging economies. *Journal of Post Keynesian Economics*, *46*(1), 113–141. DOI: 10.1080/01603477.2022.2084630

Lu, Y., Xu, R., & Ma, J. (2024). Urban Energy Transition Policy Implementation: Navigating Complex Political Realities and Overcoming Barriers to Effective Strategy Deployment. *Sustainable Cities and Society*, *101*, 105127. DOI: 10.1016/j.scs.2023.105127

Lyytinen, K., Yoo, Y., & Boland, R. J.Jr. (2016). Digital product innovation within four classes of innovation networks. *Information Systems Journal*, *26*(1), 47–75. DOI: 10.1111/isj.12093

Mandpe, A., Paliya, S., Gedam, V. V., Patel, S., Tyagi, L., & Kumar, S. (2022). Circular economy approach for sustainable solid waste management: A developing economy perspective. Waste Management & Research. *Waste Management & Research*, *41*(3), 499–511. DOI: 10.1177/0734242X221126718 PMID: 36264045

Manniche, J., Larsen, K. T., & Broegaard, R. B. (2021). The circular economy in tourism: Transition perspectives for business and research. *Scandinavian Journal of Hospitality and Tourism*, *21*(3), 247–264. DOI: 10.1080/15022250.2021.1921020

Mansouri, S. A., Nematbakhsh, E., Ahmarinejad, A., Jordehi, A. R., Javadi, M. S., & Marzband, M. (2022). A hierarchical scheduling framework for resilience enhancement of decentralized renewable-based microgrids considering proactive actions and mobile units. *Renewable & Sustainable Energy Reviews*, *168*, 112854. DOI: 10.1016/j.rser.2022.112854

Ma, R., Abid, N., Yang, S., & Ahmad, F. (2023). From crisis to resilience: Strengthening climate action in OECD countries through environmental policy and energy transition. *Environmental Science and Pollution Research International*, *30*(54), 115480–115495. DOI: 10.1007/s11356-023-29970-z PMID: 37882927

Marami, H., Tsapekos, P., Adiby, A., Birkved, M., Alvarado-Morales, M., Zhang, Z., Angelidaki, I., Khademi, S., Motevali, A., & Khoshnevisan, B. (2024). To what extent do geopolitical conditions affect the environmental sustainability of the circular economy? A comparative LCA study of lactic acid production in Denmark and Iran. *Sustainable Production and Consumption*, *47*, 363–383. DOI: 10.1016/j.spc.2024.04.006

Market Research. BIZ, (2024). Sustainable Tourism Market Report By Tourism Type (Ecotourism, Adventure Tourism, Cultural Tourism, Wildlife Tourism, Community-based Tourism, Green Tourism, Agro-tourism), By Service Type (Accommodation [Eco-friendly Hotels, Green Resorts, Sustainable Lodges], Transportation [Eco-friendly Transport, Public Transport, Bicycle Rentals], Activities and Tours, Food and Beverage, By End User, By Region and Companies - Industry Segment Outlook, Market Assessment, Competition Scenario, Trends and Forecast 2024-2033, https://marketresearch.biz/report/sustainable-tourism-market/

Marshall, S., & Pearse, R. (2025). Reading the room: Developing a practical justice politics of regional energy transition. *Environmental Politics*, *34*(1), 70–91. DOI: 10.1080/09644016.2024.2347164

Masini, F. (2021). William Nordhaus: A disputable Nobel [Prize]? Externalities, climate change, and governmental action. *European Journal of the History of Economic Thought*, *28*(6), 985–1004. DOI: 10.1080/09672567.2021.1963798

McFarlane, A., C., & Norris, F. H. (2006). Definitions and concepts in disaster research. In Norris, F. H. (Eds.), *Methods for disaster mental health research* (pp. 3–19). Guilford Press.

Meadows, D. H., Meadows, D. L., Randers, J., & Behrens, W. W. (1972). *The Limits to Growth*. Universe Books.

Mercedes-Benz Stadium. (2024). *Sustainability at Mercedes-Benz Stadium | Leaders in Zero Waste*. Www.mercedesbenzstadium.com. https://www.mercedesbenzstadium.com/sustainability

Metilelu, O. O., & Michael, O. (2022). Adeniyi, Matthew I. Ekum, (2022). Modelling the Dynamic Effect of Environmental Pollution on Coastal Tourism. *Scientific African*, *17*, e01364. DOI: 10.1016/j.sciaf.2022.e01364

Mikhno, I., Ihnatenko, N., Cherniaiev, O., Vynogradnya, V., Atstaja, D., & Koval, V. (2023). Construction waste recycling in the circular economy model, IOP Conference Series: Earth and Environmental Science, Volume 1126, 2nd International Conference on Environmental Sustainability in Natural Resources Management 31/10/2022 - 01/11/2022 Riga, Latvia, IOP Conf. Ser.: Earth Environ. Sci. 1126 012003, DOI DOI: 10.1088/1755-1315/1126/1/012003

Min, Y., & Lee, H. W. (2024). Quantifying clean energy justice: Impact of Solarize programs on rooftop solar disparities in the Pacific Northwest. *Sustainable Cities and Society*, *104*, 105287. Advance online publication. DOI: 10.1016/j.scs.2024.105287

Miranda, M. L., Hastings, D. A., Aldy, J. E., & Schlesinger, W. H. (2011). The environmental justice dimensions of climate change. *Environmental Justice*, *4*(1), 17–25. DOI: 10.1089/env.2009.0046

Miyamori, D., Yoshida, S., Kashima, S., Koike, S., Ishii, S., Okazaki, Y., Ikeda, K & Matsumoto, M. (2023). How the 2018 Japan floods impacted nursing home admissions for older persons: A longitudinal study using the long-term care insurance comprehensive database. *Journal of the American Medical Directors Association, 24(3)*, 368–375e1. DOI: 10.1016/j.jamda.2022.11.021 PMID: 36587929

Morales, R., Marín, L. G., Roje, T., Caquilpan, V., Sáez, D., & Nuñez, A. (2024). Microgrid planning based on computational intelligence methods for rural communities: A case study in the José Painecura Mapuche community, Chile. *Expert Systems with Applications*, *235*, 121179. DOI: 10.1016/j.eswa.2023.121179

Möslinger, M., Ulpiani, G., & Vetters, N. (2023). Giulia Ulpiani, Nadja Vetters, (2023). Circular economy and waste management to empower a climate-neutral urban future. *Journal of Cleaner Production*, *421*, 138454. DOI: 10.1016/j.jclepro.2023.138454

Mukhtiar, F. (2024). Comparing Microgrid Planning and Management Methods for Energy Resilience and Sustainability. *Power System Technology*, *48*(2), 1811–1831. DOI: 10.52783/pst.758

Munonye, W. C. (2025). Integrating Circular Economy Principles in Business Strategies: A Policy-Driven Approach. *Circular Economy and Sustainability*, *5*(3), 1887–1896. DOI: 10.1007/s43615-025-00523-5

National Programme. "Digital Bulgaria 2025". (2019). https://www.mtc.government.bg/en/category/85/national-program-digital-bulgaria-2025-and-road-map-its-implementation-are-adopted-cm-decision-no73005-12-2019

Nawar, K. N., Mahbub, T., Tashfiq, R. A., & Rashid, T. U. (2024). Municipal Solid Waste Collection, Transportation, and Segregation. In Environmental Engineering and Waste Management: Recent Trends and Perspectives. Springer Nature. DOI: 10.1007/978-3-031-58441-1_2

Naylor, C. (2024, August 7). Sports Event Food Waste: An Olympic-Sized Problem | BioteCH4. *Biotech*, *4*, •••. https://biotech4.co.uk/sports-event-food-waste/

Neumann, J. E., Price, J., Chinowsky, P., Wright, L., Ludwig, L., Streeter, R., Jones, R., Smith, J. B., Perkins, W., Jantarasami, L., & Martinich, J. (2015). Climate change risks to US infrastructure: Impacts on roads, bridges, coastal development, and urban drainage. *Climatic Change*, *131*(1), 97–109. DOI: 10.1007/s10584-013-1037-4

NFL PLAY 60. (2020). Www.heart.org. https://www.heart.org/en/professional/educator/nfl-play-60

Nixon, K. D., Schyns, Z. O. G., Luo, Y., Ierapetritou, M. G., Vlachos, D. G., Korley, L. S. T. J., & Epps, T. H.III. (2024). Analyses of circular solutions for advanced plastics waste recycling. *Nature Chemical Engineering*, *1*(10), 615–626. DOI: 10.1038/s44286-024-00121-6

Njangang, H., Tadadjeu, S., & Kamguia, B. (2024). From policy to progress: Do female parliamentarians influence energy justice? *Climate Policy*. Advance online publication. DOI: 10.1080/14693062.2024.2424231

Nordhaus, W. (1993). Reflections on the economics of climate *change*. *The Journal of Economic Perspectives*, *7*(4), 11–25. DOI: 10.1257/jep.7.4.11

Nordhaus, W. (2019). Climate change: The ultimate challenge for economics. *The American Economic Review, 109*(6), 1991–2014. DOI: 10.1257/aer.109.6.1991

Nsabimana, R., Perelman, S., Walheer, B., & Mapapa, M. (2024). Effectiveness and efficiency in access to reliable electricity: The case of East African countries. *Socio-Economic Planning Sciences, 93*, 101898. DOI: 10.1016/j.seps.2024.101898

Obersteiner, G., Gollnow, S., & Eriksson, M. (2021). Carbon footprint reduction potential of waste management strategies in tourism. *Environmental Development, 39*, 100617. DOI: 10.1016/j.envdev.2021.100617 PMID: 34513580

OECD. (2019). *Waste Management and the Circular Economy in Selected OECD Countries: Evidence from Environmental Performance Reviews, OECD Environmental Performance Reviews*. OECD Publishing., DOI: 10.1787/9789264309395-

OECD. (2020). *The Circular Economy in Cities and Regions: Synthesis Report, OECD Urban Studies*. OECD Publishing., DOI: 10.1787/10ac6ae4-

Olağandışı Durumlarda Sağlık Hizmetleri. (2021). *Kastamonu–Sinop sel afeti 1. ay raporu*. https://www.ttb.org.tr

Olympians' support for #Sport4Climate campaign gathers pace. (2025). Olympians.org. https://olympians.org/news/183/olympians-support-for-sport4climate-campaign-gathers-pace/

Olympics. (2024). *Carbon Neutral Games - Paris 2024*. Olympics.com. https://olympics.com/en/paris-2024/our-commitments/the-environment/carbon-method

Onu, U. G., de Doile, G. N. D., Zambroni de Souza, A. C., & Balestrassi, P. P. (2024). Economic sustainability and social inclusion in rural electrical grid design. *Utilities Policy, 89*, 101779. DOI: 10.1016/j.jup.2024.101779

Oswald, A. J., & Stern, N. (2019). *Why does the economics of climate change matter so much, and why has the engagement of economists been so weak?* https://www.andrewoswald.com/docs/ClimatechangeOswaldSternSept2019forRES.pdf

Ottenburger, S. S., Cox, R., Chowdhury, B. H., Trybushnyi, D., Omar, E. A., Kaloti, S. A., Ufer, U., Poganietz, W.-R., Liu, W., Deines, E., Müller, T. O., Möhrle, S., & Raskob, W. (2024). Sustainable urban transformations based on integrated microgrid designs. *Nature Sustainability*, *7*(8), 1067–1079. DOI: 10.1038/s41893-024-01395-7

Oxfam (2011). *Gender mainstreaming: Putting women's rights at the heart of all we do*. https://oxfamilibrary.openrepository.com

Pak, M. (2023). *Afetlerde psikososyal destek hizmetlerinin etkisi: Sinop Ayancık örneği* (Unpublished doctoral dissertation). Ankara University.

Parlikar, A., Tepe, B., Möller, M., Hesse, H., & Jossen, A. (2024). Quantifying the carbon footprint of energy storage applications with an energy system simulation framework — Energy System Network. *Energy Conversion and Management*, *304*, 118208. DOI: 10.1016/j.enconman.2024.118208

Parthenon, E. Y. (2025). Atık Yönetimi ve Geri Dönüşüm Sektör Analizi, https://www.ey.com/tr_tr/technical/ey-turkiye-yayinlar-raporlar/atik-yonetimi-geri-donusum-sektor-analizi

Patil, K. D., De, J., Patil, V. K., & Kulkarni, M. M. (2024). Environmental effects and threats of waste: Understanding threats and challenges to ecosystem, health, and sustainability and mitigation strategies. In *From Waste to Wealth*. Springer Nature., DOI: 10.1007/978-981-99-7552-5_3

Paxton, T. K., Donnellan-Fernandez, R., & Hastie, C. (2023). An exploratory study of women and midwives' perceptions of environmental waste management – homebirth as climate action. *Midwifery*, *127*, 103844. Advance online publication. DOI: 10.1016/j.midw.2023.103844 PMID: 37918131

Perkumienė, D., Atalay, A., Safaa, L., & Grigienė, J. (2023). Sustainable Waste Management for Clean and Safe Environments in the Recreation and Tourism Sector: A Case Study of Lithuania, Turkey and Morocco. *Recycling*, *8*(4), 56. DOI: 10.3390/recycling8040056

Pillay, R., Hansraj, R., & Singaram, V. S. (2024). Decision-making tree to support sustainable waste management in optometric clinical practice and healthcare training programmes. *Clinical & Experimental Optometry*. Advance online publication. DOI: 10.1080/08164622.2024.2433594 PMID: 39603595

Politt, H. (2019). *The contribution of post-Keynesian economics to climate policy and meeting global decarbonisation targets*, https://www.postkeynesian.net/downloads/events/Pollitt_2019.pdf

Popa, G., Stegaroiu, I., Robescu, V.-O., Barbu, I., & Ilie, V. (2010). Sustainable development strategy - The key of environmental and organizational management. Proceedings of the International Conference on Energy and Environment Technologies and EquipmProceedings of the International Conference on Energy and Environment Technologies and Equipment, EEETE '10, 88–92. https://www.scopus.com/inward/record.uri?eid=2-s2.0-79952635181&partnerID=40&md5=f939070b8ba44ab91e0a71900ccb9075

Precedence Research. (2025). Sustainable Tourism Market Size, Share, and Trends 2025 to 2034, Sustainable Tourism Market (By Type: Island Tourism, Mountain Tourism, Coastal Tourism; By Application: Group, Family, Solo, Couples; By Age group: Millennial, Baby Boomers, Generation X, Silver hair) - Global Industry Analysis, Size, Trends, Leading Companies, Regional Outlook, and Forecast 2025 to 2034, https://www.precedenceresearch.com/sustainable-tourism-market

Public Health England. (2020). https://assets.publishing.service.gov.uk/media/61f933b1d3bf7f78e469cd77/phe-annual-report-and-accounts-2020-to-2021-web-accessible.pdf

Puntillo, P. (2023). Circular economy business models: Towards achieving sustainable development goals in the waste management sector—Empirical evidence and theoretical implications. *Corporate Social Responsibility and Environmental Management*, *30*(2), 941–954. DOI: 10.1002/csr.2398

Ragossnig, A. M., & Schneider, D. R. (2019). Circular economy, recycling and end-of-waste. Waste Management & Research. *Waste Management & Research*, *37*(2), 109–111. DOI: 10.1177/0734242X19826776 PMID: 30803418

Ralston Fonseca, F., Craig, M., Jaramillo, P., Bergés, M., Severnini, E., Loew, A., Zhai, H., Cheng, Y., Nijssen, B., Voisin, N., & Yearsley, J. (2021). Effects of Climate Change on Capacity Expansion Decisions of an Electricity Generation Fleet in the Southeast U.S. *Environmental Science & Technology*, *55*(4), 2522–2531. DOI: 10.1021/acs.est.0c06547 PMID: 33497216

Ramasar, V., Busch, H., Brandstedt, E., & Rudus, K. (2022). When energy justice is contested: A systematic review of a decade of research on Sweden's conflicted energy landscape. *Energy Research & Social Science*, *94*, 102862. Advance online publication. DOI: 10.1016/j.erss.2022.102862

Ram, M., & Bracci, E. (2024). Waste Management, Waste Indicators and the Relationship with Sustainable Development Goals (SDGs): A Systematic Literature Review. *Sustainability (Basel)*, *16*(19), 8486. Advance online publication. DOI: 10.3390/su16198486

Reinman, S. L. (2012). Intergovernmental panel on climate change (IPCC). *Reference Reviews*, *26*(2), 41–42. DOI: 10.1108/09504121211205250

Reis, G. L., Brandao, D. I., Oliveira, J. H., Araujo, L. S., & Cardoso Filho, B. J. (2022). Case Study of Single-Controllable Microgrid: A Practical Implementation. *Energies*, *15*(17), 6400. DOI: 10.3390/en15176400

Rezai, A., Taylor, L., & Foley, D. (2013). Climate change and macroeconomic structural change. *Cambridge Journal of Economics*, *37*(5), 1207–1226.

Rodríguez, C., Florido, C., & Jacob, M. (2020). Circular Economy Contributions to the Tourism Sector: A Critical Literature Review. *Sustainability (Basel), 12*(11), 4338. DOI: 10.3390/su12114338

Rolewicz-Kalińska, A., Lelicińska-Serafin, K., & Manczarski, P. (2020). The Circular Economy and Organic Fraction of Municipal Solid Waste Recycling Strategies. *Energies, 13*(17), 4366. DOI: 10.3390/en13174366

Romero-Perdomo, F., Carvajalino-Umaña, J. D., Moreno-Gallego, J. L., Ardila, N., & González-Curbelo, M. Á. (2022). Research Trends on Climate Change and Circular Economy from a Knowledge Mapping Perspective. *Sustainability (Basel), 14*(1), 521. DOI: 10.3390/su14010521

Ross, J. (2017). Don't confuse digital with digitization. *MIT Sloan Management Review*. https://sloanreview.mit.edu/article/dont-confuse-digital-with-digitization/

Rossetto, D. (2023). The relative importance of carbon markets to the waste management sector's future contribution to climate change commitments under the Paris Agreement: Insights from Australia. *Carbon Neutrality, 2*(1), 25. Advance online publication. DOI: 10.1007/s43979-023-00063-7

Rudan, E. (2023). Circular Economy of Cultural Heritage—Possibility to Create a New Tourism Product through Adaptive Reuse. *Journal of Risk and Financial Management, 16*(3), 196. DOI: 10.3390/jrfm16030196

Saeed, A., Zafar, M. W., Manita, R., & Zahid, N. (2024). The role of audit quality in waste management behavior. *International Review of Economics & Finance, 89*, 1203–1216. DOI: 10.1016/j.iref.2023.08.019

Sandberg, J., Holmstrom, H., & Lyytinen, K. (2020). Digitization and phase transitions in platform organizing logics: Evidence from the Process Automation Industry. *Management Information Systems Quarterly, 44*(1), 129–153. DOI: 10.25300/MISQ/2020/14520

SAP. (2016). Digital business modeling: A structural approach toward digital transformation. *White Paper Digital Transformation.* DOI: 10.13140/RG.2.1.3167.3368

Sasmoko, Z., Zaman, K., Malik, M., Awan, U., Handayani, W., Jabor, M. K., & Asif, M. (2022). Environmental Effects of Bio-Waste Recycling on Industrial Circular Economy and Eco-Sustainability. *Recycling*, *7*(4), 60. DOI: 10.3390/recycling7040060

Scandinavian Waste-to-Energy Studies (2023).

Schaal, T. (2024, February 9). *Environmental Impact of the Super Bowl - OWL ESG*. OWL ESG. https://owlesg.com/2024/02/09/environmental-impact-of-the-super-bowl/

Schaltegger, S., Herzig, C., Kleibir, O., & Muller, J. (2002). *Sustainability management in business enterprises: Concepts and instruments for sustainable organization development. The Federal Ministry for the Environment, Nature Conservation and Nuclear Safety (BMU)*. Division for Environment and Economy., http://fox.leuphana.de/portal/files/1198766/German_Federal_Ministry_for_the_Environment.pdf

Schroeder, P., Anggraeni, K., & Weber, U. (2019). The Relevance of Circular Economy Practices to the Sustainable Development Goals. *Journal of Industrial Ecology*, *23*(1), 77–95. DOI: 10.1111/jiec.12732

Serafimova, V., & Vasilev, V. (2024). Digital culture as a competitive advantage in the sustainable development of organizations. *Agora International Journal of Economical Sciences.*, *18*(1), 210–222. DOI: 10.15837/aijes.v18i1.6724

Serrano, T., Aparcana, S., Bakhtiari, F., & Laurent, A. (2021). Contribution of circular economy strategies to climate change mitigation: Generic assessment methodology with focus on developing countries. *Journal of Industrial Ecology*, *25*(6), 1382–1397. DOI: 10.1111/jiec.13178

Shah, H. H., Amin, M., Pepe, F., & Tregambi, C. (2024). Sustainable waste management and waste-to-energy in the context of a circular economy through various waste management technologies. *Environmental Science and Pollution Research International*, *32*(48), 27503–27522. Advance online publication. DOI: 10.1007/s11356-024-33223-y PMID: 38589589

Shamshiry, Elmira, Nadi, Behzad, Bin Mokhtar, Mazlin, Komoo, Ibrahim, Saadiah Hashim, Halimaton, Yahaya, Nadzri, Integrated Models for Solid Waste Management in Tourism Regions: Langkawi Island, Malaysia, Journal of Environmental and Public Health, 2011, 709549, 5 pages, 2011. DOI: 10.1155/2011/709549

Sharma, S. (2023, February 17). *What is the Carbon Footprint of Sport? - The Carbon Literacy Project*. The Carbon Literacy Project - Relevant Climate Change Learning for Everyone. https://carbonliteracy.com/what-is-the-carbon-footprint-of-sport/

Shrimali, G., & Sen, V. (2020). Scaling reliable electricity access in India: A public-private partnership model. *Energy for Sustainable Development*, *55*, 69–81. DOI: 10.1016/j.esd.2020.01.006

Simran, S. (2025). How To Reduce Cost And Waste In A Hotel - Guide For Hotel Owner, https://www.bdtask.com/blog/how-to-reduce-cost-and-waste-in-a-hotel

Singh, P., Sharma, S., & Dhanorkar, M. (2023). Aquatic plant biomass-derived porous carbon: Biomaterials for sustainable waste management and climate change mitigation. *International Journal of Environmental Science and Technology*, *20*(11), 12955–12970. DOI: 10.1007/s13762-022-04601-1

Soğancı, E. (2024). *Selden etkilenen topluluklara yönelik psikososyal destek hizmetleri: Bozkurt örneği* (Unpublished doctoral dissertation). Ankara University.

Sørensen, F., & Bærenholdt, J. O. (2020). Jørgen Ole Bærenholdt, (2020). Tourist practices in the circular economy. *Annals of Tourism Research*, *85*, 103027. Advance online publication. DOI: 10.1016/j.annals.2020.103027

Sovacool, B. K., Baum, C. M., & Low, S. (2022). Climate protection or privilege? A whole systems justice milieu of twenty negative emissions and solar geoengineering technologies. *Political Geography*, *97*, 102702. Advance online publication. DOI: 10.1016/j.polgeo.2022.102702

Spash, C. L. (2020). A tale of three paradigms: Realising the revolutionary potential of ecological economics. *Ecological Economics*, *169*, 1–14. DOI: 10.1016/j.ecolecon.2019.106518

Statista Research Department. (2025). Concerns over the environmental impact of tourism among adults in the United Kingdom as of January 2025, by age, https://www.statista.com/statistics/1135573/tourism-sustainability-concerns-uk-by-age/?srsltid=AfmBOopsQ7nzfQK_Man6ttR9TJzL5MYhEHHHjWIkV0DQ0XChQ8gbnnck

Stefanova, D. P., Vasilev, V. P., & Efremovski, I. P. (2023). Re-innovative organizational design: Sustainable branding and effective communication - applied models in a world with new borders/without borders. In *Handbook of Research on Achieving Sustainable Development Goals With Sustainable Marketing*. 112-127. DOI: 10.4018/978-1-6684-8681-8.ch006

Stern, N. (2007). *The economics of climate change: The Stern review*. Cambridge University Press. DOI: 10.1017/CBO9780511817434

Stern, N., Stiglitz, J., & Taylor, C. (2022). The economics of immense risk, urgent action and radical change: Towards new approaches to the economics of climate change. *Journal of Economic Methodology*, *29*(3), 181–216. DOI: 10.1080/1350178X.2022.2040740

Stevenson, N. (2017). Ulrich Beck. In Turner, B. S., & Elliott, A. (Eds.), *Çağdaş toplum kuramından portreler* (pp. 435–449). İletişim.

Storm, S. (2021). The political economy of decarbonization: A global perspective. *Development and Change*, *52*(4), 949–970.

Strippoli, R., Gallucci, T., & Ingrao, C. (2024). Circular economy and sustainable development in the tourism sector – An overview of the truly-effective strategies and related benefits. *Heliyon*, *10*(17), e36801. https://www.cell.com/heliyon/fulltext/S2405-8440(24)12832-0. DOI: 10.1016/j.heliyon.2024.e36801 PMID: 39281586

Suárez-Eiroa, B., Fernández, E., Méndez-Martínez, G., & Soto-Oñate, D. (2019). Emilio Fernández, Gonzalo Méndez-Martínez, David Soto-Oñate, (2019). Operational principles of circular economy for sustainable development: Linking theory and practice. *Journal of Cleaner Production, 214*, 952–961. DOI: 10.1016/j.jclepro.2018.12.271

Su, J., Zhang, R., Dehghanian, P., Kapourchali, M. H., Choi, S., & Ding, Z. (2024). Renewable-dominated mobility-as-a-service framework for resilience delivery in hydrogen-accommodated microgrids. *International Journal of Electrical Power & Energy Systems, 159*, 110047. DOI: 10.1016/j.ijepes.2024.110047

Sumiyati, S., Samadikun, B. P., Widiyanti, A., Budihardjo, M. A., Al Qadar, S., & Puspita, A. S. (2024). Life cycle assessment of agricultural waste recycling for sustainable environmental impact. *Global Journal of Environmental Science and Management, 10*(2), 907–938. DOI: 10.22035/gjesm.2024.02.30

Sustainability Directory. (2025). Tourism Waste Management, https://pollution.sustainability-directory.com/term/tourism-waste-management/#:~:text=Types%20of%20Tourism%20Waste&text=Solid%20Waste%20%E2%86%92%20This%20is,%2C%20attractions%2C%20and%20public%20spaces

Sustainability» About the SDG's. (n.d.). https://sustainability.sport/about-the-sdgs

Sustainable Development Report. (2023). *The 2023 Sustainable Development Report is out.* https://sdsn.bg/en/the-2023-sustainable-development-report-is-out/

Sustainable Travel International. (2024). *Carbon footprint of tourism.* Sustainable Travel International. https://sustainabletravel.org/issues/carbon-footprint-tourism/

Tahir, W., Farhan, M., Bhatti, A. R., Butt, A. D., & Farid, G. (2024). A modified control strategy for seamless switching of virtual synchronous generator-based inverter using frequency, phase, and voltage regulation. *International Journal of Electrical Power & Energy Systems*, *157*, 109805. DOI: 10.1016/j.ijepes.2024.109805

Taşkın, A., & Demir, N. (2020). Life cycle environmental and energy impact assessment of sustainable urban municipal solid waste collection and transportation strategies. *Sustainable Cities and Society*, *61*, 102339. Advance online publication. DOI: 10.1016/j.scs.2020.102339

Tayyab, Q., Qani, N. A., Elkholy, M. H., Ahmed, S., Yona, A., & Senjyu, T. (2024). Techno-economic configuration of an optimized resident microgrid: A case study for Afghanistan. *Renewable Energy*, *224*, 120097. DOI: 10.1016/j.renene.2024.120097

Tereshchenko, E., Happonen, A., Porras, J., & Vaithilingam, C. A. (2023). Green Growth, Waste Management, and Environmental Impact Reduction Success Cases From Small and Medium Enterprises Context: A Systematic Mapping Study. *IEEE Access : Practical Innovations, Open Solutions*, *11*, 56900–56920. DOI: 10.1109/ACCESS.2023.3271972

This Earth Day, Lewis Hamilton is fighting climate change. (n.d.). Https:// Www.we.org. https://www.we.org/en-US/we-stories/we-day/this-earth-day-lewis-hamilton-is-fighting-climate-change

Thomas, L., & Hicks, W. (2024, December 10). *Can sport empower us to be more sustainable?* World Economic Forum. https://www.weforum.org/stories/2024/12/can-sport-empower-us-to-be-more-sustainable/

Thomas, R., & Healy, L. (2010). Be prepared. In Gillespie, D., & Danso, K. (Eds.), *Disaster concepts and issues* (pp. 255–270). CSWE Press.

Tisi, Y. S. A. B., Matos, F. A., & Carneiro, M. L. N. M. (2023). Development of waste-to-energy through integrated sustainable waste management: The case of ABREN WtERT Brazil towards changing status quo in Brazil. *Waste Disposal & Sustainable Energy*, *5*(3), 295–308. DOI: 10.1007/s42768-022-00127-x PMID: 36687498

Trakic, A., Karim, R., & Ahmed, P. K. (2024). Energy Justice: What is It and Why It Matters? *Sustainable Development Goals Series*, (Part F3606), 3–10. DOI: 10.1007/978-981-97-6059-6_1

Travel, W. P. (2025). Global Tourism Statistics 2024:Trends and Forecasts, https://wptravel.io/global-tourism-industry-statistics/#h-global-tourism-statistics-2019

Tsosie, R. (2007). Indigenous people and environmental justice. *University of Colorado Law Review*, *78*(4), 1625–1678.

Türk Kızılayı. (2022). *Kızılay Bozkurtlu ailelere desteğini sürdürüyor.* https://www.kizilay.org.tr/Haber/HaberArsiviDetay/6593

Türkiye Cumhuriyeti Cumhurbaşkanlığı Strateji Ve Bütçe Başkanlığı. (2023). Katı Atık Yönetimi, On İkinci Kalkınma Planı 2024-2028, Özel İhtisas Komisyonu Raporu, Ankara, 2023. https://www.sbb.gov.tr/wp-content/uploads/2025/08/Kati-Atik-Yonetimi-Ozel-Ihtisas-Komisyon-Raporu_01082025.pdf

Türkiye, U. N. P. D. (2021). The Zero Waste Project receives UNDP Turkey's first Global Goals Action Award, https://www.undp.org/turkiye/press-releases/zero-waste-project-receives-undp-turkeys-first-global-goals-action-award

TURKSTAT. (2025). Atık İstatistikleri, 2024, https://data.tuik.gov.tr/Bulten/Index?p=Atik-Istatistikleri-2024-54134

TURKSTAT. (2025a), Manufacturing industry waste management statistics, 2022, 2024 (Ton -Tonnes), Atık İstatistikleri, 2024, https://data.tuik.gov.tr/Bulten/Index?p=Atik-Istatistikleri-2024-54134

TURKSTAT. (2025b). Waste statistics of manufacturing industry sub-sectors, 2022, 2024 (Ton-Tonnes), Atık İstatistikleri, 2024, https://data.tuik.gov.tr/Bulten/Index?p=Atik-Istatistikleri-2024-54134

TURKSTAT. (2025c). Municipal sewage sludge statistics, 2022, 2024, Atık İstatistikleri, 2024, https://data.tuik.gov.tr/Bulten/Index?p=Atik-Istatistikleri-2024-54134

TURKSTAT. (2025d). Municipal waste management statistics, 2024, Atık İstatistikleri, 2024, https://data.tuik.gov.tr/Bulten/Index?p=Atik-Istatistikleri-2024-5413

TURKSTAT. (2025e). Municipal waste services statistics, 2024, Atık İstatistikleri, 2024, https://data.tuik.gov.tr/Bulten/Index?p=Atik-Istatistikleri-2024-54134

TURKSTAT. (2025f). Mining establishments waste statistics, 2022, 2024 (Ton -Tonnes), Atık İstatistikleri, 2024, https://data.tuik.gov.tr/Bulten/Index?p=Atik-Istatistikleri-2024-54134

Turner, B. S. (2006). *Sociology*. Cambridge University Press.

UNDRR. (2023). 2022 Global Natural Disaster Asessment Report. https://www.preventionweb.net/publication/2022-global-natural-disaster-assessment-report

United Nations (2015). *General Assembly resolution adopted on 3 June 2015*. https://www.preventionweb.net/files/resolutions/N1516716.pdf?startDownload=true

United Nations (2022). "Global Waste Management Outlook."

United Nations. (1987). *Report of the World Commission on environment and development: Our Common Future*. http://www.un-documents.net/our-common-future.pdf

United Nations. (2015). Paris Agreement. In report of the conference of the parties to the United Nations Framework Convention on Climate Change (21st Session, 2015: Paris).

UNPD. (2023). What is circular economy and why does it matter? https://climatepromise.undp.org/news-and-stories/what-is-circular-economy-and-how-it-helps-fight-climate-change

Van Caneghem, J., Van Acker, K., De Greef, J., Wauters, G., & Vandecasteele, C. (2019). Waste-to-energy is compatible and complementary with recycling in the circular economy. *Clean Technologies and Environmental Policy*, *21*(5), 925–939. DOI: 10.1007/s10098-019-01686-0

Varjani, S., Haq, I., Pandey, A., Gupta, V. K., & Bui, X.-T. (2024). Waste Management in Climate Change and Sustainability Perspectives: Organic and Medical Waste. In *Waste Management in Climate Change and Sustainability Perspectives: Organic and Medical Waste*. CRC Press., DOI: 10.1201/9781003386902

Vilá, O., Cutts, B., Whitney, K., & Rivers, L. (2023). Environmental justice in disaster recovery: Recognition of the Latinx community by nonprofit leaders. *Climate Risk Management, 40*. DOI: 10.1016/j.crm.2023.100502

Viva! (2021). Viva! Novak Djokovic. https://viva.org.uk/lifestyle/living-vegan/celebrities/sportspeople/novak-djokovic/

Vivas, M. A. U., Amórtegui, L. S., & Guerrero-García-Rojas, H. (2025). Understanding the characteristics of the green, circular and bioeconomy models within the context of sustainable development. *Energy, Sustainability and Society, 15*(1), 27. DOI: 10.1186/s13705-025-00528-w

Wade, M. (2015). Digital business transformation. A conceptual framework. *Global Center for Digital Business Transformation.* An IMD and Cisco Initiative. https://www.imd.org/uupload/IMD.WebSite/DBT/Digital%20Business%20Transformation%20Framework.pdf

Wang, K. C. M., Lee, K. E., & Mokhtar, M. (2021). Solid Waste Management in Small Tourism Islands: An Evolutionary Governance Approach. *Sustainability (Basel), 13*(11), 5896. DOI: 10.3390/su13115896

Whalen, C. J. (2023). Minsky meets Kapp: A post-Keynesian institutionalist approach to addressing climate change. *Journal of Economic Issues, 57*(2), 598–605. DOI: 10.1080/00213624.2023.2202560

Wilby, R. L., Orr, M., Depledge, D., Giulianotti, R., Havenith, G., Kenyon, J. A., Matthews, T. K. R., Mears, S. A., Mullan, D. J., & Taylor, L. (2022). The impacts of sport emissions on climate: Measurement, mitigation, and making a difference. *Annals of the New York Academy of Sciences, 1519*(1), 20–33. DOI: 10.1111/nyas.14925 PMID: 36377356

Wilson, D. C. (2023). Learning from the past to plan for the future: An historical review of the evolution of waste and resource management 1970–2020 and reflections on priorities 2020–2030 – The perspective of an involved witness. *Waste Management & Research*, *41*(12), 1754–1813. DOI: 10.1177/0734242X231178025 PMID: 37732707

World Bank. (2019). *Adaptive social protection and disaster risk management: A case study in Japan*. https://openknowledge.worldbank.org

World Economic Forum. (2018). *This is why Japan's floods have been so deadly*. https://www.weforum.org/stories/2018/07/japan-hit-by-worst-weather-disaster-in-decades-why-did-so-many-die/

worldrugby.org. (n.d.). *Environmental Sustainability Plan 2030 | World Rugby*. Www.world.rugby. https://www.world.rugby/organisation/sustainability/environment

Xu, A., Wang, C., Tang, D., & Ye, W. (2022). Tourism circular economy: Identification and measurement of tourism industry ecologization. *Ecological Indicators*, *144*, 109476. DOI: 10.1016/j.ecolind.2022.109476

Xu, J., & Xu, W. (2024). Financing sustainable smart city Projects: Public-Private partnerships and green Bonds. *Sustainable Energy Technologies and Assessments*, *64*, 103699. DOI: 10.1016/j.seta.2024.103699

Yalçın, G. (2020). *Toplumsal cinsiyet perspektifinden doğal afetler* (Unpublished master's thesis). Hacettepe University.

Yang, M., Chen, L., Wang, J., Msigwa, G., Osman, A. I., Fawzy, S., Rooney, D. W., & Yap, P.-S. (2023). Circular economy strategies for combating climate change and other environmental issues. *Environmental Chemistry Letters*, *21*(1), 55–80. DOI: 10.1007/s10311-022-01499-6

Yap, C. C., Wu, S. L., Soon, P. V., Berezina, E., Aroua, M. K., & Gew, L. T. (2024). Culinary waste management for a healthier planet: A qualitative study. *Cogent Social Sciences*, *10*(1), 2388178. Advance online publication. DOI: 10.1080/23311886.2024.2388178

Yeşilyurt, M. (2023). *Jandarmanın doğal afetlerdeki önemi: Bozkurt sel felaketi örneği* (Unpublished master's thesis). Kastamonu University.

Yildirim, S. (2022). The Consumer Role for Sustainable Development: How Consumers Contribute Sustainable Development Goals. In I. Management Association (Ed.), Research Anthology on Measuring and Achieving Sustainable Development Goals (pp. 872-888). IGI Global Scientific Publishing. DOI: 10.4018/978-1-6684-3885-5.ch047

Yıldırım, D. Ç., Ektiren, B., & Erdoğan, F. (2023). Burak Ektiren, Fatma Erdoğan, (2023). The threshold effect of tourism, growth, and population interactions on environmental pollution in Mediterranean countries. *Tourism Management Perspectives*, *47*, 101118. DOI: 10.1016/j.tmp.2023.101118

Yıldırım, S. (2025). Unveiling Circularity: Challenges and Barriers in Transitioning to Circular Practices. In Singh, P., Daga, S., Yadav, K., & Mishra, V. (Eds.), *Circular Economy and Environmental Resilience*. Springer., DOI: 10.1007/978-3-031-93091-1_8

Yildirim, S., & Bostanci, S. H. (2025). Informal Recycling Sector in the Developing World and Circular Economy: Case of Türkiye. In Borthakur, A., & Singh, P. (Eds.), *Circular Economy and Green Transition in the Global South*. Springer., DOI: 10.1007/978-3-031-90827-9_3

Yıldırım, S., & Kantarcı, T. (2022). A Review on Sustainability Policies of Businesses: Recycling and Waste Reduction. *Journal of Recycling Economy & Sustainability Policy*, *1*(1), 1–9. https://respjournal.com/index.php/pub/article/view/1

Yıldırım, S., Sevik, N., Kandpal, V., & Yıldırım, D. C. (2024). The Role of Green Brands on Achieving 2030 Sustainable Development Goals (2030 SDGs). In Kankaew, K., Chaudhary, S., & Widtayakornbundit, S. (Eds.), *Contemporary Management and Global Leadership for Sustainability* (pp. 141–162). IGI Global Scientific Publishing., DOI: 10.4018/979-8-3693-1273-5.ch009

Yıldırım, S., & Vasilev, V. (2024). The future of green collar workers: Green Transition in employment. In Yıldırım, S., Yıldırım, D., Demirtaş, I., & Kandpal, V. (Eds.), *Green Transition Impacts on the Economy* (pp. 1–16). Society, and Environment., DOI: 10.4018/979-8-3693-3985-5.ch001

Yıldırım, S., & Yıldırım, D. Ç. (2025). Adapting Smart Waste Management and Recycling Practices Among the Circular Economy. In Kandpal, V., Gunasekaran, A., Jaswal, A., & Mukherjee, D. (Eds.), *Rethinking Resources. Approaches to Global Sustainability, Markets, and Governance.* Springer., DOI: 10.1007/978-981-96-9055-8_18

Yıldırım, S., Yıldırım, D. Ç., Aydın, K., & Erdoğan, F. (2021). Regime-dependent effect of tourism on carbon emissions in the Mediterranean countries. *Environmental Science and Pollution Research International, 28*(39), 54766–54780. DOI: 10.1007/s11356-021-14391-7 PMID: 34014485

Yoshida, S., Kashima, S., Okazaki, Y., & Matsumoto, M. (2023). Effects of 2018 Japan floods on healthcare costs and service utilization in Japan: A retrospective cohort study. *BMC Public Health, 23*(1), 1-10. DOI: 10.1186/s12889-023-15205-w PMID: 36755264

Yuan, X., Suvarna, M., Lim, J. Y., Pérez-Ramírez, J., Wang, X., & Ok, Y. S. (2024). Active Learning-Based Guided Synthesis of Engineered Biochar for CO2 Capture. *Environmental Science & Technology, 58*(15), 6628–6636. DOI: 10.1021/acs.est.3c10922 PMID: 38497595

Yuleva-Chuchulayna, R. (2024). Models for assessment of competitive positions and the competitiveness of companies. *Journal of Management Sciences and Applications, 2*(2), 232–244. https://www.unwe.bg/doi/JOMSA/2024.2/JOMSA.2024.2.06.pdf. DOI: 10.37075/JOMSA.2024.2.06

Yung, J. (2023, April 12). *Sport can support sustainability, contribute to environment-related SDGs.* Sportanddev. https://www.sportanddev.org/latest/news/sport-can-support-sustainability-contribute-environment-related-sdgs

Yusoff, M. S.. (2022). Solid Waste Management in the Tourism Industry. In Wang, L. K., Wang, M. H. S., & Hung, Y. T. (Eds.), *Solid Waste Engineering and Management. Handbook of Environmental Engineering* (Vol. 25). Springer., DOI: 10.1007/978-3-030-96989-9_1

Zakeri, B. (2024). Digitalization for Resilient and Sustainable Energy Transitions. *Energies, 17*(21), 5434. DOI: 10.3390/en17215434

Zero Waste Foundation. (n.d). https://sifiratikvakfi.org/en

Zhang, L., & Gao, J. (2016). Exploring the effects of international tourism on China's economic growth, energy consumption and environmental pollution: Evidence from a regional panel analysis. *Renewable & Sustainable Energy Reviews, 53*, 225–234. DOI: 10.1016/j.rser.2015.08.040

Zhang, Y., Cheng, C., Yang, T., Jin, X., Jia, Z., Shen, J., & Wu, X. (2022). Assessment of climate change impacts on the hydro-wind-solar energy supply system. *Renewable & Sustainable Energy Reviews, 162*, 112480. DOI: 10.1016/j.rser.2022.112480

Zheng, L., Huang, B., Qiu, H., & Bai, H. (2024). 'The role of social media followers' agency in influencer marketing: A study based on the heuristic–systematic model of information processing'. *International Journal of Advertising, 43*(3), 554–579. DOI: 10.1080/02650487.2023.2229148

Zhong, F., Tian, J., Zhao, C., Zha, S., Chen, X., & Zhang, Y. (2024). Assessing energy justice in climate change policies: An empirical examination of China's energy transition. *Climate Policy, 24*(3), 362–377. DOI: 10.1080/14693062.2023.2261894

Zhou, Z., & Zhang, L. (2022). Sustainable waste management and waste to energy: Valuation of energy potential of MSW in the Greater Bay Area of China. *Energy Policy, 163*, 112857. Advance online publication. DOI: 10.1016/j.enpol.2022.112857

About the Contributors

Seda Bostancı, who holds two associate professorship degrees, has a primary area of expertise in the field of Local Governments, Urban and Environmental Policies. She has been continuing his academic career as an associate professor at Tekirdağ Namık Kemal University, Faculty of Economics and Administrative Sciences, Department of Political Science and Public Administration since 2016. She served as the founding Department Head of the Department of Political Science and Public Administration between 2016-2022. She graduated from Mimar Sinan Fine Arts University with a BA in 2000 and an MA in 2002. She completed her doctorate at Istanbul Technical University in 2008. She received a BA in Public Administration from Anadolu University. She worked at Sarıyer Municipality between 2004-2010. She continued his academic career at Okan University between 2010-2016. She has been involved in various academic activities in England, Hungary and Bulgaria. She has 4 SSCI publications and her main research areas are local governments, urbanization and environmental issues, political ecology, climate and water policies. Bostancı, received the Emerald Outstanding Paper award for her article in 2022.

Seda Yıldırım hava a PhD degree in marketing. She currently works at Tekirdag Namık Kemal University in Turkey. She studies sustainable development, 2030 sustainable development goals, sustainable consumption patterns, higher education, consumption patterns, environmental quality, and food security. She has many publications including book chapters, proceedings, and articles (in journals indexed with WOS and SCOPUS).

She won the Emerald literati award- Outstanding Paper Award 2022. #LiteratiAwards

Valentin Vasilev is the head of the "Center for Leadership and Public Policy" and a Professor of "National Security" ("Crisis situations and human resources management") at the Higher School of Security and Economy /HSSI/, located in the city of Plovdiv, Bulgaria. In the period from 2000 to 2021, he was part of the South-West University "Neofit Rilski", Blagoevgrad, being the deputy dean of the Faculty of Law and History from 2010 to 2019, as well as the head of the "National Security and Public Administration" department/2018/. At the university, he went through all the academic positions – assistant, senior assistant, chief assistant, associate professor. He is the author of more than a hundred publications in scientific publications in the areas of human resources management, crises, teamwork, motivation and organizational behavior, public policies, and more.

***

İdris Akkuzu is a faculty member at the Department of Political Science and Public Administration at Tekirdağ Namık Kemal University. His research interests include political economy, migration, ecology, and sports.

Shivam Bhartiya is an expert with more than 8 years of experience in Tourism Management. Prior to taking up the current role at the CDOE, Jain (deemed-to-be) University, Bengaluru, Karnataka, Dr. Bharitya serverd at Central University of Karnataka, GD Goenka University, and Hemvati Nandan Bahuguan Garhwal Central University as an Assistant Professor at Hemvati Nandan Bahuguan Garhwal Central University. He holds a PhD in Tourism Management with expertise in community development, destination planning and management, tourism and gender, tourism marketing and tourism psychology.

Ramratan Guru is an Assistant Professor Level III with the School of Design (SOD) at Mody University, India. His research interests include new materialism, sustainability in fashion and textiles; circular fashion;

waste management; technical textile, green composite materials and ethical fashion. Dr. Ramratan has authored six books, published over 170 high-impact journal and conference papers, and contributed to more than 15 book chapters. He holds five granted patents and has filed more than 25 patent National and International applications. Dr. Ramratan has developed over eight innovative cellulosic fibers, earning recognition for his research in various countries, including Canada, the USA, Europe, the UK, and Brazil. He actively collaborates with universities across Brazil, Europe, Canada, the USA, and Asia. Currently, he is pursuing postdoctoral research at the School of Engineering, Department of Textile Engineering, University of Minho, Portugal. He has extensive publications in high impact journals as well as books and book chapters.

Viktoriya Kalaydzhieva is a Doctor of Economics and Chief Assistant Professor in the Department of "Business Management and Marketing" in the Faculty of Economics of South-West University "Neofit Rilski", Blagoevgrad, Bulgaria. She obtained her PhD degree at South-West University "Neofit Rilski" in 2015. Viktoriya Kalaydzhieva conducts lectures and seminars in the fields of Innovation Management, Advertising Policy, Marketing Planning and Entrepreneurial Strategies. Her research interests are in the broad areas of innovations, competitiveness, entrepreneurship, modern management, digitalization and digital transformation of business, sustainable development, green entrepreneurship, etc. She has published two books, two book chapters and 27 papers in scientific journals and conferences.

Arvinder Kaur is working as an Assistant Professor at the University School of Business, Chandigarh University, in the field of Finance since 2016. She has done Masters in Finance and Control with distinction Masters and graduation in commerce. She is a qualified member of Institute of Company Secretaries of India. Also, have a license as a distributor of Mutual Funds under NISM. She is currently pursuing Ph.D from Punjabi University, Patiala. She has seven years of academic experience and expertise in teaching specialized subjects like Taxation, Financial Accounting, Corporate Accounting and GST. She has 20 research papers presented (national and international conferences) and published in various renowned journals and 7 book chapters. She attended

16 faculty Development Programs and workshops in her tenure. Her area of Interest is Fintech.

Pawan Kumar, PhD, is a prominent academic and researcher with over 20 years of expertise in business studies. A professor at Graphic Era Deemed to be University, Dehradun, India, he has supervised six PhD scholars and published over 60 research papers, including 40 in SCOPUS/ Web of Science, and 14 books, including eight indexed in Scopus. His works, featured by prominent publishers like Emerald, IGI Global, Nova Science, and Cambridge Scholars, have gained recognition. Dr. Kumar earned his PhD in Commerce from Kurukshetra University in 2007 and holds an M.Phil. and M.Com. Degrees. Specializing in finance, banking, and securities markets, he has authored books on green finance, AI, and fintech. His innovative research includes five patents and has served as a guest editor for Scopus-indexed journals. In 2022, he received the prestigious Researcher Award from Chandigarh University. Dr. Kumar's impactful work on sustainable finance, blockchain, and fintech is globally recognized. His active participation in high-impact conferences and faculty development programs reflects his commitment to advancing research, teaching, and innovation in commerce and management.

M. Rajkumar is a highly experienced academician, with distinctive knowledge in business research techniques, marketing research, organizational behavior, and human capital management. At the center of his research is a persistent interest in leadership dynamics, motivation science, and organizational citizenship—threads he has pursued through multiple publications in Scopus-indexed journals and presentations given at global conferences. Aside from mentoring students and program management, Dr. Rajkumar is a peer reviewer for a leading academic journal (ABDC). He is also the Associate Editor for The IUJ Journal of Management. His proficiency in research methodologies, data analysis (SPSS, AMOS, SmartPLS), and structural equation modeling has been instrumental in conducting numerous workshops and faculty development programs

Kavita Mathad, a Gold Medallist in her post-graduation from Karnatak University, Karnataka, India, and an alumnus of Harvard

Business School Online holds rich academic and research experience. She has conducted several International Conferences, Business Conclaves, Webinars, and FDPs/MDPs and has actively participated in various webinars and FDPs MDPs as a resource person. To her credit, she has a good number of Scopus and ABDC indexed research papers and book publications, Major Research Projects, and consultancy done mainly in International Business, Digital Economy, and Sustainability. Dr Kavita Mathad has worked at the most eminent institutes and Universities Like XIME, Alliance University, IFIM Business School, CHRIST University, GIBS Business School, Presidency University, in Bangalore and Galgotias University, NCR, in various leadership positions and as a member of the core team of accreditations/rankings of NAAC, NBA and the QS Ranking. Dr. Kavita is known for her leadership acumen in building new perspectives and values in the HEIs ecosystem.

Amitava Pal is an accomplished academician, mentor, and researcher specializing in Human Resource Management and Sports Management. He is currently pursuing a PhD from ICFAI University and holds a Post Graduate Program from XLRI, Jamshedpur, along with a Bachelor of Engineering in Computer Science (First Class with Distinction). He heads the academics of International Institute of Sports & Management (IISM), Mumbai, where he oversees academic delivery, resource management, and innovation initiatives(IISM received the prestigious President's Award from Ramnath Kovindji for contributing in the development of sports in India). He also chairs the Innovation and Entrepreneurship Development Centre at IISM. A recognized mentor and advisor, he worked with Niti Aayog's Atal Innovation Mission, Startup India Hub, University of Chicago's IIC Fellows, and IIT Bombay's Abyuday. Amitava has an extensive research portfolio, including co-authoring sports case studies, impact reports, and industry white papers. His contributions range from guiding SOPs for major Khelo India to National Games . His expertise spans sports analytics, youth engagement, fitness trends, and sports infrastructure.

Kushal Singh is an esteemed academician and industry professional, currently serving as the Head of Department (i/c), Department of Tourism Administration, Tripura University (A Central University). He holds a Ph.D. in Tourism Management from Mahatma Gandhi Chitrakoot Gramodaya Vishwavidyalaya, India's first rural university, along with a Master's in Tourism Administration and a B.Com. from Himachal Pradesh University, Shimla. With over a decade of academic experience and three years in the tourism and hospitality industry, he has previously worked as an Assistant Professor in the Department of Aviation Hospitality and Travel Management at Gujarat University, Ahmedabad, and contributed to institutions like CT University, Ludhiana; Dev Samaj College for Women, Ferozepur; and Government Degree College, Sujanpur Tira. A prolific researcher and nationalist thinker, Dr. Singh has published 32+ research articles, authored nine books, presented 40 research papers at national and international conferences, and written 42 newspaper articles. He also holds three patents and is highly active on platforms like ORCID, Web of Science, Google Scholar, and ResearchGate. His research spans rural tourism, medical tourism, sustainable tourism, aviation management, and hospitality. As a mentor, he has successfully guided six Ph.D. scholars and continues to inspire future researchers. Known for his vision of tourism as a transformative tool for empowerment, sustainability, and national development, Dr. Singh remains dedicated to advancing knowledge and shaping future leaders in the tourism sector. His teaching interests span a wide range of topics, including the Business of Tourism, Tourism Resources, Destination Planning, Airlines Management and Airport Operation, Introduction to Cabin Crew, Hospitality Law, and Event Management.

Durmuş Çağrı Yıldırım works at Tekirdağ Namık Kemal University in Türkiye. He is an expert in economics. He studies on sustainable development, climate change, energy policies, education and health policies in general. He has many book chapter and book in IGI Global publishing. He has several publications in journal indexed by Web of Science and SCOPUS.

Index

L

O

P

R

S

T

W

Z